# Mechanical Properties and Microstructure of Forged Steel

# Mechanical Properties and Microstructure of Forged Steel

Editors

**Andrea Di Schino**
**Koh-ichi Sugimoto**

MDPI • Basel • Beijing • Wuhan • Barcelona • Belgrade • Manchester • Tokyo • Cluj • Tianjin

*Editors*
Andrea Di Schino
Università di Perugia
Italy

Koh-ichi Sugimoto
Shinshu University
Japan

*Editorial Office*
MDPI
St. Alban-Anlage 66
4052 Basel, Switzerland

This is a reprint of articles from the Special Issue published online in the open access journal *Metals* (ISSN 2075-4701) (available at: https://www.mdpi.com/journal/metals/special_issues/forged_steel).

For citation purposes, cite each article independently as indicated on the article page online and as indicated below:

LastName, A.A.; LastName, B.B.; LastName, C.C. Article Title. *Journal Name* **Year**, *Volume Number*, Page Range.

**ISBN 978-3-0365-6362-6 (Hbk)**
**ISBN 978-3-0365-6363-3 (PDF)**

# Contents

# About the Editors

**Andrea Di Schino**

Andrea Di Schino received he Physics Degree from the University of Pisa, Italy, in 1996 and a Ph.D. in Materials Engineering from the University of Naples in 2000. He received the National Habilitation for a Full Professorship in 2016. After more than 15 years spent in Centro Sviluppo Materiali Spa, he joined the University of Perugia as an Associate Professor of Metallurgy. Today he is Head of the Industrial Engineering School.

**Koh-ichi Sugimoto**

Koh-ichi Sugimoto recieved his Ph.D in Mechanical Engineering from Tokyo Metropolitan University in 1985. After a year spent in Daido Steel, he joined Shinshu University as an Associate Professor and a Full Professor of Mechanical Engineering. Today he is an Emeritus Professor of Shinshu University.

# Preface to "Mechanical Properties and Microstructure of Forged Steel"

Forged steels represent an interesting material family, both from a scientific and commercial point of view, due to their many applications. Based on this, it is essential to deeply understand the relationships between properties and microstructure and how to drive them through processing. Despite their diffusion as a consolidated material, many research fields are actively employing new applications. At the same time, new innovations are arising from the manufacturing processing of such materials, including the possibility to manufacture them from metal powders suitable for 3D printing. This Special Issue embraces the interdisciplinary work covering physical metallurgy and processing, reporting the experimental and theoretical progress concerning microstructural evolution during processing, and microstructure–properties relations.

**Andrea Di Schino and Koh-ichi Sugimoto**

*Editors*

*Editorial*

# Mechanical Properties and Microstructure of Forged Steels

Andrea Di Schino

Dipartimento di Ingegneria, Università Degli Studi di Perugia, via G. Duranti 93, 06125 Perugia, Italy; andrea.dischino@unipg.it

**Citation:** Di Schino, A. Mechanical Properties and Microstructure of Forged Steels. *Metals* **2021**, *11*, 1177. https://doi.org/10.3390/met11081177

Received: 25 June 2021
Accepted: 23 July 2021
Published: 24 July 2021

**Publisher's Note:** MDPI stays neutral with regard to jurisdictional claims in published maps and institutional affiliations.

Forged steels represent a quite interesting material family, both from a scientific and commercial point of view, following many applications they can be devoted to [1]. Following from that, it is, therefore, essential to deeply understand the relations between properties and the microstructure, and how to drive them by process. Despite their diffusion as a consolidated material, many research fields are active regarding new applications. At the same time, innovations are coming from the manufacturing process of such a family of materials. In this framework in particular, the role of heat treatments in obtaining even complex microstructures is still quite an open matter, also thanks to the design of innovative heat treatments [2,3] devoted to forged components [4–7]. The Special Issue scope embraces interdisciplinary work covering physical metallurgy and processes, reporting about the experimental and theoretical progress concerning microstructural evolution during the processing and microstructure-properties relations.

The book collects manuscripts from academic and industrial researchers with stimulating new ideas and original results. The present book consists of three research papers.

M. Algarni in his research analyzes the mechanical properties and fracture behavior of two cold-work tool steels: AISI "D2" and "O1". Tool steels are an economical and efficient solution for manufacturers due to their superior mechanical properties. The demand for tool steels is increasing yearly due to the growth in the transportation production around the world. Nevertheless, AISI "D2" and "O1" (locally made) tool steels behave differently due to the varying content of their alloying elements. There is also a lack of information regarding their mechanical properties and behavior. Therefore, this study aims to investigate the plasticity and ductile fracture behavior of "D2" and "O1" via several experimental tests. The tool steels' behavior under monotonic quasi-static tensile and compression tests was analyzed. The results of the experimental work showed different plasticity behavior and ductile fracture among the two tool steels. Before fracture, clear necking appeared on the "O1" tool steel, whereas no signs of necking occurred on the "D2" tool steel. In addition, the fracture surface of the "O1" tool steel showed a cup–cone fracture mode, and the "D2" tool steel showed a flat-surface fracture mode. The dimple-like structures in scanning electron microscope (SEM) images revealed that both tool steels had a ductile fracture mode [8].

L. Pezzato et al. describe the effects of a second tempering treatment on the microstructural properties and impact toughness of a structural steel EN 10025-6 S690. The steel was first forged and quenched in water after austenitization at 890 °C for 4 h. After quenching, different tempering treatments were performed, at 590 °C in single or multiple steps. The effect of these treatments was evaluated both in microstructural terms—by means of optical microscopy scanning, transmission electron microscopy and X-ray diffraction—and in terms of impact toughness. The mechanical behavior was correlated with the microstructure and a remarkable increase in impact toughness was found after the second tempering treatment due to carbide shape change [9].

S. Mancini et al. propose a new approach in order to describe the plastic strain at the core of the piece. FEM takes into account the plastic deformation at the core of the forged pieces. At the first stage, a thermomechanical FEM model was implemented in the MSC Marc commercial code in order to simulate the open die forging process. Starting from the results obtained through FEM simulations, a set of equations describing the plastic strain at the core of the piece were identified depending on forging parameters (such as the length of the contact surface between the tools and ingot, tool's connection radius, and reduction in the piece height after the forging pass). An Artificial Neural Network (ANN) was trained and tested in order to correlate the equation coefficients with the forging to obtain the behavior of the plastic strain at the core of the piece [10].

**Funding:** This research received no external funding.

**Acknowledgments:** As Guest Editors, I would like to especially thank Harley Wang, Assistant Editor, for his support and his active role in the publication. I am also grateful to the entire staff of the *Metals* Editorial Office for the precious collaboration. Last but not least, we express our gratitude to all the contributing authors and reviewers: without your excellent work it would not have been possible to accomplish this Special Issue that we hope will be a piece of interesting reading and reference literature.



## References

1. Di Schino, A. Manufacturing and application of stainless steels. *Metals* **2020**, *10*, 327. [CrossRef]
2. Di Schino, A.; Testani, C. Heat treatment of steels. *Metals* **2021**, *11*, 1168. [CrossRef]
3. Mancini, S.; Langellotto, L.; Di Nunzio, P.E.; Zitelli, C.; Di Schino, A. Defect reduction and quality optimization by modeling plastic deformation and metallurgical evolution in ferritic stainless steels. *Metals* **2020**, *10*, 186. [CrossRef]
4. Di Schino, A.; Alleva, L.; Guagnelli, M. Microstructure evolution during quenching and tempering of martensite in a medium C steel. *Mat. Sci. Forum* **2012**, *715–716*, 860–865. [CrossRef]
5. Di Schino, A.; Di Nunzio, P.E. Metallurgical aspects related to contact fatigue phenomena in steels for back up rolling. *Acta Metall. Slovaca* **2017**, *23*, 62–71. [CrossRef]
6. Di Schino, A. Analysis of phase transformation in high strength low alloyed steels. *Metalurgija* **2017**, *56*, 349–352.
7. Di Schino, A.; Di Nunzio, P.E.; Turconi, G.L. Microstructure evolution during tempering of martensite in a medium C steel. *Mat. Sci. Forum* **2007**, *558–559*, 1435–1441. [CrossRef]
8. Algarni, M. Mechanical properties and microstructure characterization of AISI "D2" and "O1" cold work tool steels. *Metals* **2019**, *9*, 1169. [CrossRef]
9. Pezzato, L.; Gennari, C.; Chukin, D.; Toldo, M.; Sella, F.; Toniolo, M.; Zambon, A.; Brunelli, K.; Dabalà, M. Study of the effect of multiple tempering on the impact toughness of forged S690 structural steel. *Metals* **2020**, *10*, 507. [CrossRef]
10. Mancini, S.; Langellotto, L.; Zangari, G.; Maccaglia, R.; Di Schino, A. Optimization of open die ironing process through artificial neural network for rapid process simulation. *Metals* **2020**, *10*, 1397. [CrossRef]

*Review*

# Recent Progress of Low and Medium-Carbon Advanced Martensitic Steels

**Koh-ichi Sugimoto**

Department of Mechanical Systems Engineering, School of Science and Technology, Shinshu University, Nagano 380-8553, Japan; sugimot@shinshu-u.ac.jp; Tel.: +81-90-9667-4482

**Abstract:** This article introduces the microstructural and mechanical properties of low and medium-carbon advanced martensitic steels (AMSs) subjected to heat-treatment, hot- and warm- working, and/or case-hardening processes. The AMSs developed for sheet and wire rod products have a tensile strength higher than 1.5 GPa, good cold-formability, superior toughness and fatigue strength, and delayed fracture strength due to a mixture of martensite and retained austenite, compared with the conventional martensitic steels. In addition, the hot- and warm-stamping and forging contribute to enhance the mechanical properties of the AMSs due to grain refining and the improvement of retained austenite characteristics. The case-hardening process (fine particle peening and vacuum carburization) is effective to further increase the fatigue strength.

**Keywords:** advanced martensitic steel; retained austenite characteristics; microstructure; mechanical properties; heat treatment; hot-stamping; hot-forging; case hardening

**Citation:** Sugimoto, K.-i. Recent Progress of Low and Medium-Carbon Advanced Martensitic Steels. *Metals* **2021**, *11*, 652. https://doi.org/10.3390/met11040652

Academic Editor: Marcello Cabibbo

Received: 30 March 2021
Accepted: 10 April 2021
Published: 17 April 2021

**Publisher's Note:** MDPI stays neutral with regard to jurisdictional claims in published maps and institutional affiliations.



## 1. Introduction

The strain-induced transformation of austenite to martensite enhances the ductility of austenitic steels such as Fe-Ni, Fe-Ni-C, and Fe-Cr-Ni steels. These high-alloy austenitic steels are called TRansformation-Induced Plasticity (TRIP) steels [1,2]. In the 1980s, low and medium carbon Si-Mn ferritic steels subjected to intercritical annealing and then isothermal transformation (IT) or austempering process were developed by Sakuma et al. [3,4]. The steel is named low alloy TRIP-aided steel or TRIP-assisted steel because it achieves high ductility by the TRIP effect of metastable retained austenite of 5 to 30 vol %. The TRIP-aided steel was mainly applied to the automotive body parts that need high cold press formability and weldability [4–6]. Up to now, various kinds of low and medium carbon advanced ultrahigh- and high-strength steels (AHSSs) with metastable retained austenite of different volume fraction, stability, size, morphology, and chemical composition were developed for the weight reduction and the improvement of crash safety of the automotive body [7–10].

In general, the AHSSs are categorized as the following: first-, second- and third-generation AHSSs [7–9]. The second-generation AHSSs are high Mn austenitic steels with an Mn content higher than 14 mass % and are named TWinning-Induced Plasticity (TWIP) steels [11]. The first- and third-generation AHSSs except for medium Mn (MMn) steels with 4 to 12% Mn are lean micro-alloyed Si/Al-Mn steels. The third-generation AHSSs are classified into two types, Type A and Type B, by the kind of matrix structure.

(I).    First-generation AHSS: ferrite–martensite dual-phase (DP) steel [7,9,12–16], TRIP-aided polygonal ferrite (TPF) steel [3–7,9,17–19], TRIP-aided annealed martensite (TAM) steel [20–22], and complex-phase (CP) steel [7,9,15,23],

(II).   Second-generation AHSS: high Mn TWIP and TWIP/TRIP steels [7,9,11,24–28],

(III).  Third-generation AHSS (Type A): TRIP-aided bainitic ferrite (TBF) steel [9,10,29–33], one-step and two-step quenched and partitioned (Q&P) steels [7–9,25,34–41], carbide-free bainitic (CFB) steel [42–49], and duplex type medium manganese (D-MMn) steel [9,25,50–57].

(IV). Third-generation AHSS (Type B): TRIP-aided martensitic (TM) steel [58–63] and martensite-type medium manganese (M-MMn) steel [53,64–69], which are called advanced martensitic steel (AMS).

The product of tensile strength and total elongation (TS×TEl) of various AHSSs as a function of austenite or retained austenite fraction is shown in Figure 1. For the third-generation AHSSs (Type A), a TS×TEl higher than 30 GPa% is required to apply to the automotive body frame members, the seat frame members, the concrete mixer truck cylinders, etc. In connection with this, the IT process at low temperatures during martensite-start temperature ($M_s$) and martensite-finish temperature ($M_f$) is recently applied to TBF, one-step Q&P, and CFB steels [30–32,34–41,43,47,48,58–61], which achieve high tensile strength and mechanical properties due to bainitic ferrite/martensite (BF/M) structure matrix. To obtain the tensile strength higher than 1.5 GPa, TM and M-MMn steels with martensitic structure matrix are recently developed [58–69]. These steels are classified as the third-generation AHSSs (Type B). Hereafter, these third-generation AHSSs (type B) are also called low and medium-carbon "*Advanced Martensitic Steel* (AMS)", because the martensitic structure is the main matrix structure.

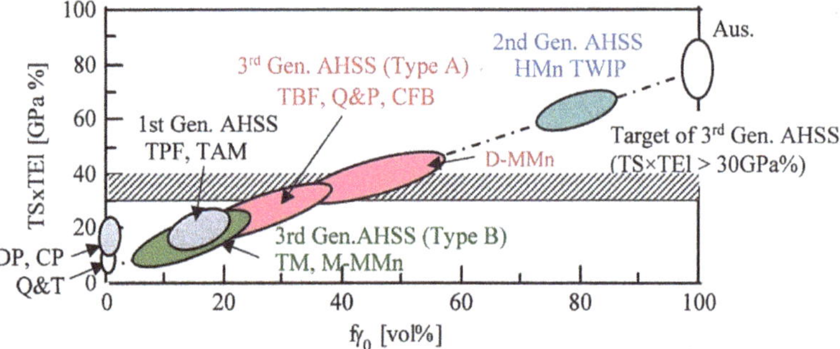


**Figure 1.** Relationship between the product of tensile strength and total elongation (TS×TEl) and initial volume fraction of austenite or retained austenite ($f\gamma_0$) in the first-, second-, and third-generation (Type A and Type B) advanced high-strength steels (AHSSs). Q&T: conventional quenched and tempered martensitic steel, DP: ferrite–martensite dual-phase steel, CP: complex-phase steel, TPF, TAM, TBF, and TM: transformation-induced plasticity (TRIP)-aided steels with polygonal ferrite, annealed martensite, bainitic ferrite, and martensite matrix structure, respectively. Q&P: one-step and two-step quenched and partitioned steel, CFB: carbide-free bainitic steel, D-MMn: duplex-type medium Mn steel, M-MMn: martensite-type medium Mn steel, HMn TWIP: high manganese TWIP steel, Aus: austenitic steel. This figure is reproduced based on Ref. [52]. Reprinted with permission from Elsevier: Mater. Sci. Eng. A, Copyright 2021.

To produce the AMS, two kinds of heat-treatment process are proposed: (1) IT process below $M_f$ [59–63] and (2) direct quenching to room temperature (DQ) [58–63,70–73]. In the case of $M_f$ > room temperature, the partitioning process will be added after the DQ process. The AMS exhibits higher TS×TEl [59–63] (Figure 1) and higher formability [60–63] than the conventional quenched and tempered (Q&T) martensitic steel. In addition, the AMS possesses excellent toughness [53,64,65,74], high fatigue strength (especially high notch-fatigue strength) [75], and high delayed fracture strength [76]. So, the AMS is expected to be applied to not only press forming sheet products but also bar-forging products such as gear, screw, etc.

This paper introduces the microstructural and mechanical properties of various low and medium-carbon AMSs. In addition, the improvement mechanisms of the mechanical properties are detailed by relating to the matrix structure, the strain-induced transformation behavior of metastable retained austenite, and/or the martensite/austenite constituent (MA phase).

## 2. Two Kinds of Heat-Treatment Process for AMSs

The heat-treatment process of AMS is as simple as the Q&T process of the conventional martensitic steels. In addition, the process after austenitizing is conducted at relatively low temperature, compared with those of TBF, Q&P, and CFB steels [58–63,66–73]. This means that quenching in an oil bath can be used for the heat treatment, not a salt bath. Regarding alloying elements of the AMS, Si, and/or Al higher than 0.5 mass% are added to suppress the carbide precipitation and increase the volume fraction of carbon-enriched retained austenite [59–63]. In some cases, micro-alloying elements of Mn, Cr, Ni, Mo, B, etc. are added to increase the hardenability of the steels [62,65]. Furthermore, V, Ti, and/or Nb are added to refine the prior austenitic grain and to increase the strength through their carbide precipitates [27,59–63,67,68].

When the $M_f$ of the AMS is higher than room temperature, the IT process below $M_f$ [59–63] or DQ process [53,58–63,70–73] immediately after austenitizing and hot-rolling is conducted using an oil bath below 200 °C (Figure 2a). In the case of the DQ process, partitioning is added after the DQ process (named DQ-P process) [59,60,70–72]. The partitioning temperatures just below and above $M_f$ are recommended to minimize the increase in carbide precipitation and the decrease in retained austenite fraction [59,60]. Unlike this, Gao et al. propose the partitioning (tempering) temperature above $M_s$ [70–72]. Such IT and DQ-P processes (Figure 2a) are applied to Si/Al-Mn steels with low and medium carbon content [59–63] and with a medium manganese content of about 5% [65,66]. Regarding M-MMn steels, air cooling to room temperature is possible instead of quenching in oil or water bath due to the high hardenability [64,77–80].

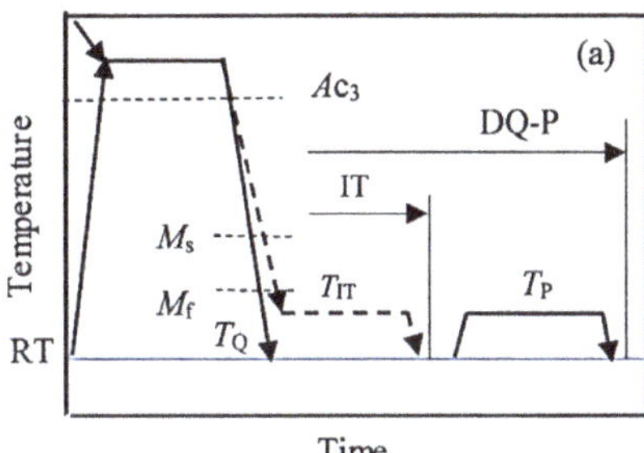


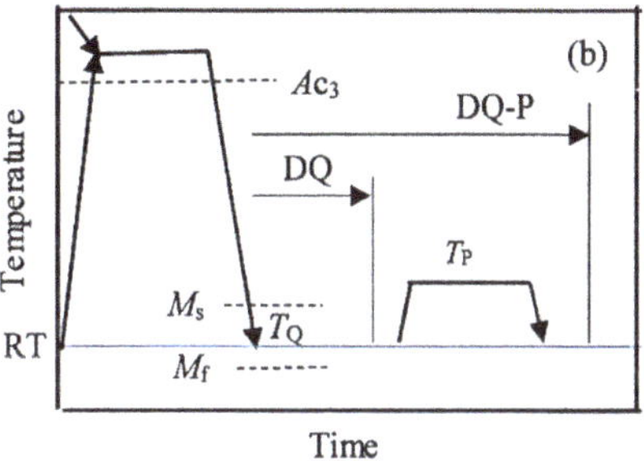


**Figure 2.** Heat treatment diagrams of (**a**) the isothermal transformation (IT) process below $M_f$ and direct quenching to room temperature (RT) followed by partitioning (DQ-P) process in a case of RT < $M_f$ [58–63] and (**b**) the DQ and DQ-P process in a case of RT > $M_f$ [67–69]. RT: room temperature, $T_{IT}$: isothermal transformation temperature, $T_Q$: quenching temperature, $T_P$: partitioning temperature.

On the other hand, when the $M_f$ of the AMS is lower than room temperature such as for M-MMn steel with a relatively high Mn content of about 10 mass %, the DQ or DQ-P process is carried out [67–69] (Figure 2b). For the DQ-P process, He et al. [67] adopt the partitioning (tempering) at 300 to 400 °C for 10 min in 0.2%C–10%Mn–2%Al–0.1%V steel.

To optimize the microstructure and mechanical properties of the AMS, ausforming at temperatures between $Ac_3$ and $M_s$ is carried out before the IT, DQ, or DQ-P process [68,81–83], in the same way as TBF [84] and CFB steels [49,85].

## 3. Microstructure and Retained Austenite Characteristics of AMSs

### 3.1. IT and DQ-P Processes (RT < $M_f$)

When 0.21%C–1.49%Si–1.50%Mn–1.0%Cr–0.05%Mn TM steel is subjected to the IT process at temperatures below $M_f$ (261 °C) and the DQ process, most of the austenite start to transform to soft coarse lath-martensite accompanied with auto-tempering [59–63] (Figures 3 and 4a). The other austenites are retained as filmy and blocky morphology in the lath-martensite structure matrix, and a part of retained austenite stays in the fine MA phase [35,59–63]. It is noteworthy that fine lath/twin-martensite in the MA phase is very hard because of fine morphology, high dislocation density, and high carbon concentration. As mentioned in the previous section, partitioning after the DQ process rises the mechanical stability of the retained austenite due to carbon-enrichment. In addition, it plays an important role in the softening of the coarse lath-martensite and fine lath/twin-martensite and carbon enrichment of retained austenite, with a small increase in carbide fraction and a small decrease in retained austenite fraction [60], as well as the relaxation of transformation strain. Such the AMSs are corresponding to low and medium-carbon TM steels [32,59–63] and M-MMn steels with relatively low Mn content [65,66]. Note that carbides precipitate only in soft coarse lath-martensite (Figure 3c). De Cooman et al. show that the carbide is transition carbide or cementite [17].

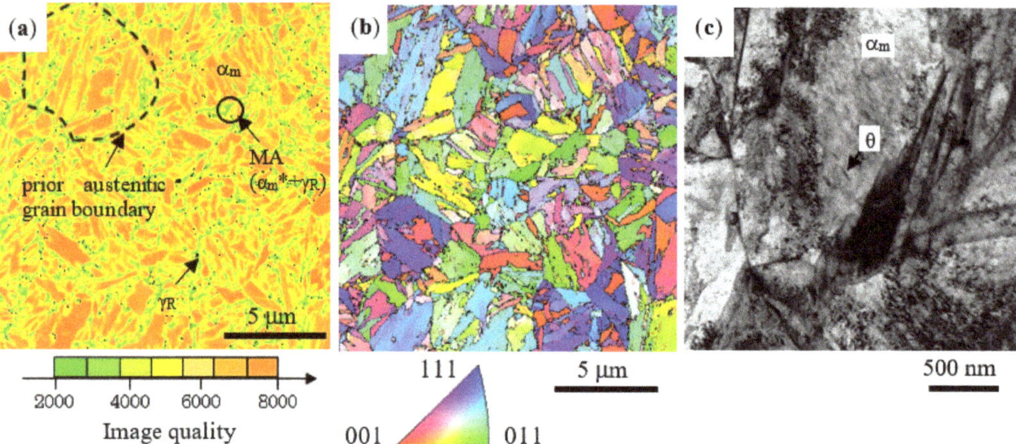


**Figure 3.** (**a**) Image quality distribution map and (**b**) inverse pole figure map of Fe-α (body centered cubic structure) phase and (**c**) TEM image in 0.21%C–1.49%Si–1.50%Mn–1.0%Cr–0.05%Mn TM steel subjected to the IT process at $T_p$ = 200 °C (< $M_f$) for 1000 s [62]. $\alpha_m$: coarse lath-martensite, $\alpha_m$*: fine lath/twin-martensite, $\gamma_R$: retained austenite (black dots), MA: martensite/austenite phase, θ: carbide in coarse lath-martensite. Reprinted with permission from ISIJ: ISIJ Int, Copyright 2021.

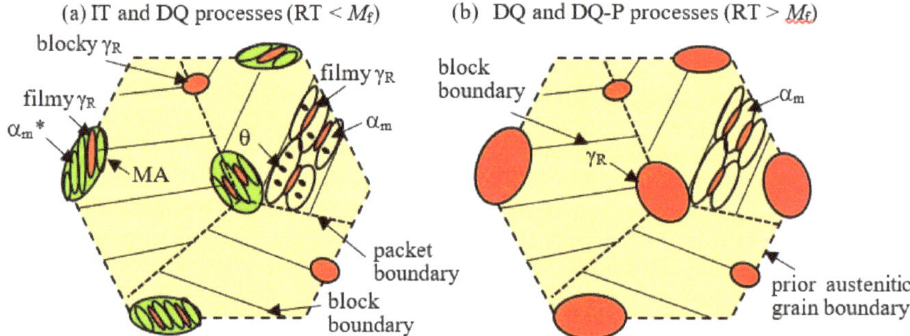


**Figure 4.** Illustration of typical microstructures of advanced martensitic steels (AMSs) subjected to (**a**) the IT and DQ processes in the case of RT < $M_f$ and (**b**) the DQ and DQ-P processes in the case of RT > $M_f$. RT: room temperature. $\alpha_m$, $\alpha_m$*, $\gamma_R$, θ and MA represent coarse lath-martensite, fine lath/twin-martensite, retained austenite, carbide and MA phase, respectively.

According to Sugimoto et al. [60,61,63], the retained austenite fraction increases with increasing IT temperature, with a constant carbon concentration, in an IT temperature range of 25 to 250 °C in 0.21%C–1.49%Si–1.50%Mn–1.0%Cr–0.05%Nb TM steel (Figure 5a), where the IT process at 25 °C is corresponding to the DQ process. It is noteworthy that the MA phase fraction increases and the carbide fraction decreases with increasing IT temperature in the TM steel (Figure 5b). The carbide fraction is 1/4 to 1/2 times that of JIS-SCM420 Q&T steel. Additions of Cr and Mo hardly influence the retained austenite fraction (about 4 vol %) but increase the MA phase fraction and decrease the carbide fraction [59,62]. An increase in Mn content significantly increases the volume fraction of retained austenite in 0.2%C–1.5%Si–1.5%Mn TM steel [65,66]. The mechanical stability of the retained austenite defined by the following strain-induced transformation factor $k$ [6] is nearly constant in an IT temperature range below $M_f$ in the same way as the carbon concentration of the retained austenite (Figure 5b,c).

$$\log f\gamma = \log f\gamma_0 - k\,\varepsilon \tag{1}$$

where $f\gamma$ is the volume fraction of retained austenite after being subjected to a plastic strain $\varepsilon$ and $f\gamma_0$ is the initial volume fraction of retained austenite. The $k$-value is higher than that of TBF steel [29,30] owing to the lower carbon concentration of the retained austenite and higher flow stress. Mn addition in 0.2%C–1.5%Si–1.5%Mn steel increases the volume fraction and mechanical stability of retained austenite because Mn is austenite former element [65,66].

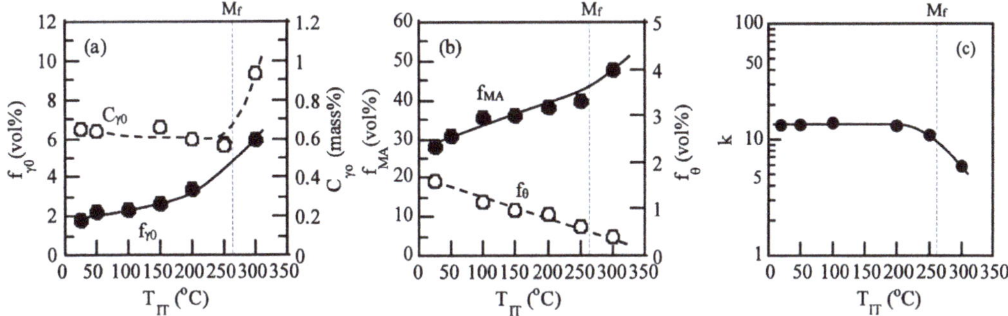

**Figure 5.** Variations in (**a**) initial volume fraction ($f\gamma_0$) and initial carbon concentration ($C\gamma_0$) of retained austenite, (**b**) volume fractions of MA phase ($f_{MA}$) and carbide ($f_\theta$), and (**c**) strain-induced transformation factor ($k$) as a function of IT temperature ($T_{IT}$) in 0.21%C–1.49%Si–1.50%Mn–1.0%Cr–0.05%Nb TM steel [61]. The IT process at 25 °C is corresponding to the DQ process. Reprinted with permission from AIST: AIST 2013, Copyright 2021.

### 3.2. DQ and DQ-P Processes (RT > $M_f$)

The DQ process at the temperatures above $M_f$ produces a simple duplex structure of soft coarse lath-martensite structure matrix and blocky and filmy retained austenite in 0.2%C–10%Mn–2%Al–0.1%V [67,68] (Figure 4b). The soft lath-martensite fraction ($f\alpha_m$) can be estimated by the following empirical equation proposed by Koistinen and Marburger [8,86].

$$f\alpha_m = 1 - \exp\left\{-1.1 \times 10^{-2}\,(M_s - T_Q)\right\} \tag{2}$$

where $T_Q$ is the quenching temperature. It is noteworthy that the retained austenite fraction is much higher than that of TM steel subjected to the IT process of Figure 2a [67,68]. Differing from the IT process (Figure 4a), the MA phase and carbide are hardly formed because the process is not cooled to temperatures below $M_f$, similar to TBF, Q&P, and CFB steels with a BF/M matrix structure [29–32,34,42–44,47]. In addition, such a microstructure without carbide in the soft lath-martensite resembles that of 0.23%C–2.3%Mn–1.5%Si–12.5%Cr–0.03%Ti–0.05%Nb martensitic stainless steel containing retained austenite [87].

According to Du et al. [69], in 0.24C–7.5Mn–1.1Si–0.1V M-MMn steel subjected to the DQ-P process, partitioning above $M_s$ softens the lath-martensite matrix structure due to the decreased carbon concentration and dislocation density, in the same way as TM steel subjected to the DQ-P process. On the contrary, the retained austenite is carbon-enriched.

Sub-cooling in liquid nitrogen decreases the retained austenite fraction. It is very important to know that the subsequent partitioning increases the volume fraction of retained austenite through the migration of austenite/martensite interface, with an increase in the carbon concentration. The AMS subjected to the DQ-P process is corresponding to M-MMn steels with $M_f$ lower than room temperature.

### 3.3. Ausforming

Ausforming at temperatures between $Ac_3$ and $M_s$ before the IT, DQ, and DQ-P process can not only reduce the $M_s$ due to the strengthening of austenite but also introduce deformation defects and elongate the microstructure [81], in the same way as TBF [84] and CFB [49] steels. If the ausforming is conducted under the conditions of relatively high temperature and large plastic strain, prior austenitic grain is refined, although dislocation density is decreased. According to He et al. [68], warm rolling at 600 °C before the DQ process is expected to avoid dynamic recrystallization and minimize dislocation recovery in 0.2%C–10%Mn–2%Al–0.1%V M-MMn steel. In this case, it is also essential to keep a finishing rolling temperature higher than the $M_s$ temperature (about 130 °C) to avoid martensitic transformation during the warm rolling process [68]. Hojo et al. [82,83] show that ausforming at 650 °C and subsequent IT process at 200 °C refines the microstructure and increases the retained austenite fraction and its stability in 0.23%C–1.5%Si–1.5%Mn–1.0%Cr–0.2%Mo–0.002%B TM steel.

## 4. Tensile Properties and Cold Formability of AMS Sheets

### 4.1. Tensile Properties

The IT process brings on the continuous yielding and low yield stress (or 0.2% offset proof stress) at room temperature in 0.21%C–1.49%Si–1.50%Mn–1.0%Cr–0.05%Nb TM steel (Figure 6). The main origin is due to a large amount of hard MA phase [61–63], which creates a preferential yielding zone in a matrix as fresh martensite second phase in dual-phase steel, as well as the strain-induced transformation of retained austenite [61–63,88].

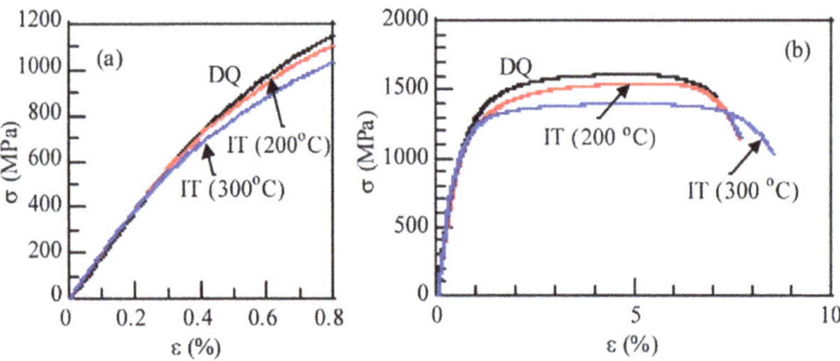


**Figure 6.** Typical engineering stress—strain ($\sigma-\varepsilon$) curves at room temperature in 0.21%C–1.49%Si–1.50%Mn–1.0%Cr–0.05%Nb TM steel subjected to the DQ process and the IT process at $T_p$ = 200 °C and 300 °C [63]. (**a**): initial stage, (**b**) overall stage [63]. Measured $M_s$ and $M_f$ of the steel are 406 °C and 261 °C, respectively.

When 0.21C–1.49Si–1.50Mn–1.0Cr–0.05Nb TM steel is subjected to the DQ process or the IT process at temperatures below 200 °C (< $M_f$), the tensile strength exceeds 1.5 GPa, although the tensile strength and yield stress slightly decrease with increasing IT temperature [60,61,63] (Figure 6). The uniform and total elongations and reduction of the area tend to slightly increase with increasing IT temperature. Partitioning after the DQ process increases the yield stress due to the relaxation of the internal residual stress, despite the softening of coarse and fine martensites and the coarsening of the carbides. In addition, the partitioning slightly decreases the uniform and total elongations [60]. Resultantly, the TM steel subjected to the DQ-P process exhibits a TS×TEl of about 11 GPa%. Torizuka and

Hanamura [53] show that 0.1%C–2%Si–5%Mn M-MMn steel subjected to the DQ process completes a TS×TEl of 30 GPa%, almost equal to those of D-MMn steel. He et al. [67] report that 0.2%C–10%Mn–2%Al–0.1%V M-MMn steel subjected to the DQ-P process ($T_p$ = 300 °C) achieves higher TS×TEl than press hardening steel and DP steel. Gao et al. find that slow cooling on the DQ process increases the total elongation with no change in tensile strength, compared to rapid cooling such as quenching in water [70].

*4.2. Formabilities*

The best combination of tensile strength and formabilities such as stretch-formability, stretch-flangeability, and bendability can be achieved by the IT process at $T_{IT}$ = 200 °C in 0.21%C–1.49%Si–1.50%Mn–1.0%Cr–0.05%Nb TM steel (Figure 7) [61]. This optimum IT temperature (200 °C) is equivalent to $M_f$—60 °C in the TM steel. All formabilities of the TM steel are superior to those of 22MnB5 Q&T steel and 0.082%C–0.88%Si–2.0%Mn DP steel. The increased stretch-formability may be caused by the TRIP effect of metastable retained austenite. Meanwhile, the high stretch-flangeability is brought from a small degree of damage to the punched hole-surface with a long shear section and a small number of tiny cracks or voids, resulting from the plastic relaxation by the strain-induced transformation of retained austenite. Such small punching damage leads to difficult crack propagation and void growth on hole-expanding [60,61]. Good bendability is considered to be caused by a high localized ductility. Partitioning after the DQ process further increases these formabilities, but the formabilities are lower than those of the IT process at 200 °C [60].

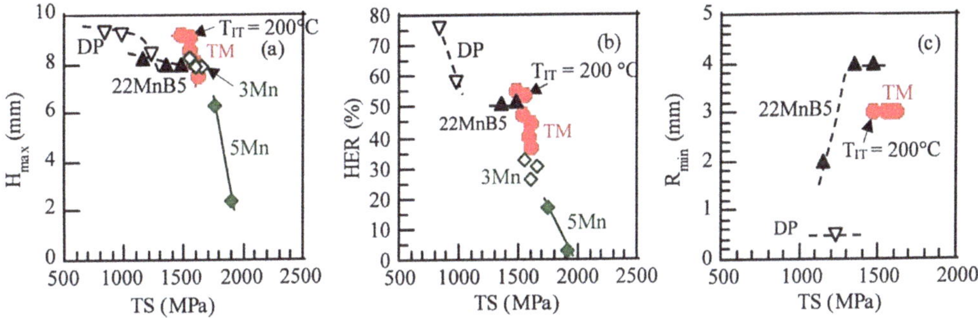


**Figure 7.** Relationships between (**a**) maximum stretch height ($H_{max}$), (**b**) hole expansion ratio (HER) and (**c**) minimum bending radius ($R_{min}$), and tensile strength (TS) in 0.21%C–1.49%Si–1.50%Mn–1.0%Cr–0.05%Nb TM steel subjected to the DQ process or the IT process at $T_{IT}$ = 100 to 250 °C for 1000 s (●). DP (▽): 0.082%C–0.88%Si–2.0%Mn ferrite–martensite dual-phase steel, DIN-22MnB5 (▲): 0.23%C–0.19%Si–1.29%Mn–0.21%Cr–0.003%B Q&T steel [61], 3Mn (◇): 0.2%C–1.5%Si–3%Mn M-MMn steel subjected to the DQ process and the IT process at $T_{IT}$ = 200 °C, 5Mn (◆): 0.2%C–1.5%Si–5%Mn M-MMn steel subjected to the DQ process and the IT process at 100 °C [66]. This figure was reproduced from Refs. [61,66]. Ref. [61] is reprinted with permission from AIST: AIST 2013, Copyright 2021.

Sugimoto and Tanino [66] show that 0.2%C-1.5%Si-5%Mn M-MMn steel subjected to the IT process at 100 °C ($<M_f$) achieves a maximum stretch height ($H_{max}$) of 6.3 mm and a hole expanding ratio (HER) of 16.7% at maximum (Figure 7). However, 0.2%C-1.5%Si-3%Mn M-MMn steel with low Mn content exhibits higher $H_{max}$ and HER, although the HER is lower than that of 22MnB5 Q&T steel [89]. Sugimoto et al. propose that a large amount of retained austenite transformed at an early stage in the M-MMn steels contradicts the increase in the stretch-formability and stretch-flangeability through the easy void initiation and growth [66]. Up to now, there is no data of bendability in the M-MMn steel.

## 5. Mechanical Properties of Heat-Treated AMS Plates and Bars

### 5.1. Impact Toughness and Fracture Toughness

According to Kobayashi et al. [74], 0.2%C-1.5%Si-1.5%Mn-(0–1.0)%Cr-(0–0.2)%Mo TM steels subjected to the DQ-P process ($T_p$ = 300 °C) possess as high Charpy V-notch impact value ($E_v$) at room temperature as TBF steels with the same chemical composition (Figure 8a). The $E_v$s and the product of TS and $E_v$ (TS×$E_v$) of 0.2%C-1.5%Si-(3, 5)%Mn M-MMn steels subjected to the DQ or the IT process decrease compared with those of TM steels, especially in 5%Mn steel [65]. Ductile-brittle transition temperatures (DBTTs) of the TM steels are far lower than those of JIS-SCM420 Q&T and TBF steels [74,90] (Figure 8b). The DBTTs of the M-MMn steels are almost the same as those of 0.2%C-1.5%Si-1.5%Mn-(0–1.0)%Cr-(0–0.2)%Mn TBF steels but are far higher than the TM steels [65].

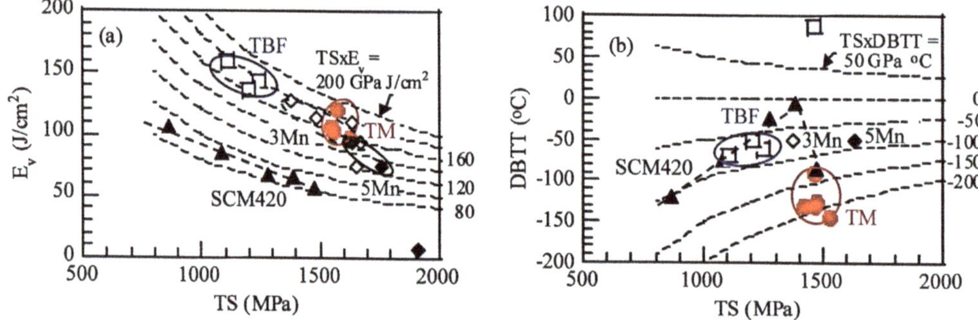


**Figure 8.** (**a**) Relationships between (**a**) Charpy V-notch impact value ($E_v$) at 25 °C and (**b**) ductile–brittle fracture transition temperature (DBTT) and tensile strength (TS) in 0.2%C–1.5%Si–1.5%Mn–(0–1.0)%Cr–(0–0.2)%Mn TBF ($T_{IT}$ = 400 °C, □) and TM (DQ-P, $T_p$ = 300 °C, •) steels, JIS-SCM420 Q&T steel ($T_T$ = 200 °C to 600 °C, ▲) and 0.2%C–1.5%Si–3%Mn (◇) and 0.2%C–1.5%Si–5%Mn (◆) M-MMn steels. This figure was reproduced from refs. [65,74]. Ref. [74] was reprinted with permission from Springer Nature: Metall. Mater. Trans. A, Copyright 2021.

Figure 9 illustrates (a) the ductile fracture (void initiation, growth and void-coalescence) and (b) brittle fracture (cleavage crack initiation and propagation) behavior of the TM steel appeared on impact tests [74,90]. In the ductile fracture, most of the voids originate at the interface of the large MA phase and the matrix structure. In this case, retained austenite suppresses the void initiation through the plastic relaxation by the strain-induced martensite transformation. As the deformation progresses further, these voids grow and coalesce (Figure 9a). Therefore, (i) a large amount of MA phase and (ii) the plastic relaxation mainly contribute to the high $E_v$s of the TM steels. Gao et al. show that 0.20%C–2.0%Mn–1.5%Si– 0.5%Cr–0.28%Mo and 0.4%C–2.0%Mn–1.7%Si–0.4%Cr TM steels subjected to the DQ-P process ($T_p$ = 280 °C > $M_s$) exhibit higher $E_v$s than TBF and CFB steels, although the $E_v$s of these steels are lower than that of two-step Q&P steels with the same chemical composition [71,72].

In the brittle fracture, the strain-induced transformation of retained austenite also suppresses the cleavage crack initiation, and the MA phase retards the crack growth (Figure 9b). Resultantly, the retained austenite and MA phase play a role in lowering the DBTT of the TM steel [74]. On the other hand, high DBTTs of 0.2%C–1.5%Si–(3,5)%Mn M-MMn steels are mainly associated with high solute Mn concentration in the matrix, although high Mn addition increases the volume fraction and mechanical stability of retained austenite and decreases the unit path of quasi-cleavage crack ($L_c$ in Figure 9b) [65].

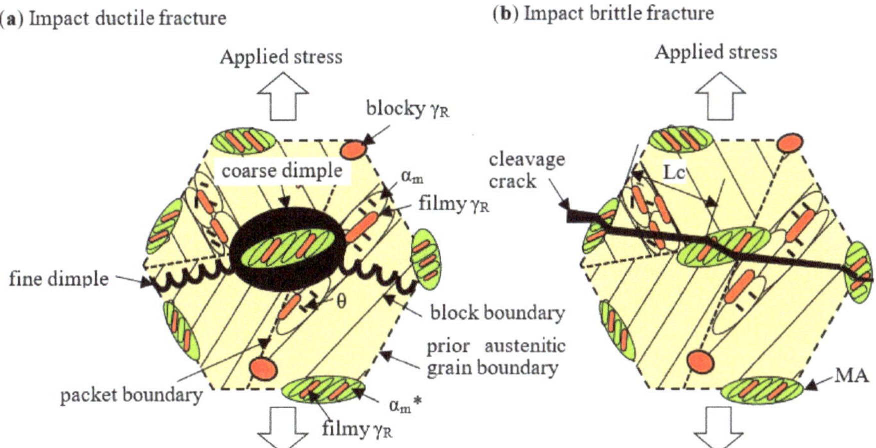



**Figure 9.** Illustrations showing (**a**) ductile fracture and (**b**) brittle fracture of TM steel appeared on impact tests [74]. *Lc*: Quasi-cleavage length affected by the MA phase located on prior austenitic, packet, and block boundaries. $\alpha_m$, $\alpha_m{}^*$, $\gamma_R$, $\theta$, and MA represent coarse lath-martensite, fine lath/twin-martensite, retained austenite, carbide, and MA phase, respectively. Reprinted with permission from Springer Nature: Metall. Mater. Trans. A, Copyright 2021.

Kobayashi et al. find that 0.2%C–1.5%Si–1.5%Mn–1.0%Cr–0.05%Nb TM steel subjected to the DQ-P process ($T_p$ = 200 to 450 °C) [91] and the IT process ($T_{IT}$ = 200 °C and 250 °C, <$M_f$) [92] shows extremely high fracture toughness ($K_{IC}$ = 132–163 MPa m$^{1/2}$) (Figure 10). The fracture toughness is two times that ($K_{IC}$ = 57–63 MPa m$^{1/2}$) of JIS-SCM420 Q&T steel tempered at $T_T$ = 200 to 400 °C and is the same degree as that of 18Ni maraging steel. It is considered that the superior fracture toughness is essentially due to (i) a large amount of MA phase and (ii) plastic relaxation by the strain-induced transformation of the retained austenite in the same way as the above impact toughness, which suppresses crack formation, growth, and coalescence as well as cleavage fracture at the pre-crack tip. Wu et al. [93] report that fracture toughness of 0.20%C–1.42%Si–1.87%Mn steel subjected to the DQ-P process ($T_p$ = 300 °C) is only $K_{IC}$ = 54.5 MPa m$^{1/2}$. The low fracture toughness significantly differs from the result of Figure 10. This result should be discussed in the future.

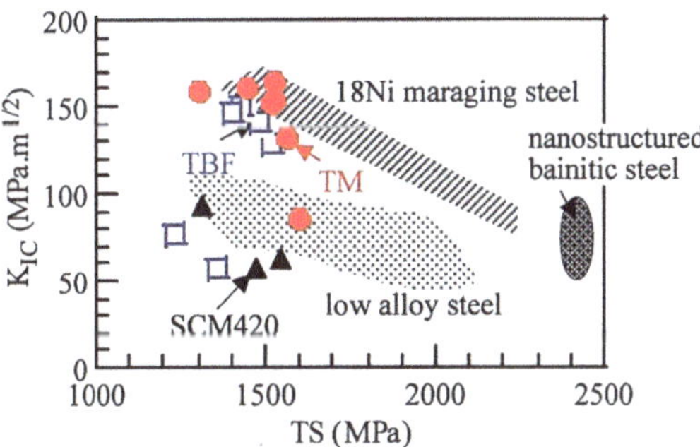



**Figure 10.** Relationship between fracture toughness ($K_{IC}$) and tensile strength (TS) [92] in 0.2%C–1.5%Si–1.5%Mn–(0–1.0)%Cr–(0–0.2)%Mn TBF (□) and TM (DQ-P, $T_p$ = 200 °C to 450 °C, •) steels and JIS-SCM420 Q&T steel ($T_T$ = 200 °C to 400 °C, ▲), compared with low alloy Q&T, 18Ni maraging and nanostructured bainitic steels [92,94]. This figure is reproduced based on Ref. [92]. Reprinted with permission from ISIJ: ISIJ Int, Copyright 2021.

*5.2. Fatigue Property*

0.2%C–1.5%Si–1.5%Mn–1.0%Cr–0.2%Mo TM steel subjected to the DQ-P ($T_p$ = 200 °C) process exhibits large fatigue hardening despite a tensile strength over 1.5 GPa [95] (Figure 11a), in the same way as TBF [96], CFB [97], and high alloy TRIP (16%Cr–7%Mn–8%Ni and 16%Cr–6%Mn–6%Ni) [98] steels. Conversely, conventional martensitic steel such as JIS-SCM420 Q&T steel exhibits fatigue softening.

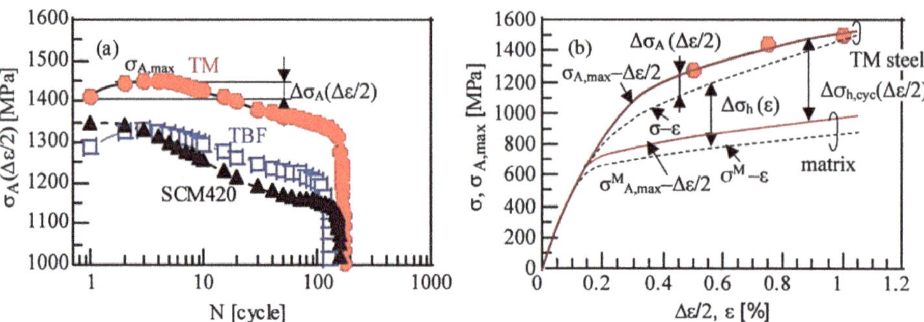

**Figure 11.** (**a**) Variations in stress amplitude ($\sigma_A(\Delta\varepsilon/2)$) as a function of the number of cycles ($N$) under total strain amplitude of $\Delta\varepsilon$ = 1.5% for 0.2%C–1.5%Si–1.5%Mn–1.0%Cr–0.2%Mo TM subjected to the IT process at $T_{IT}$ = 200 °C and TBF steel ($T_{IT}$ = 350 °C), and JIS-SCM420 Q&T steel ($T_T$ = 200 °C). (**b**) Monotonic stress–strain ($\sigma$-$\varepsilon$: dotted lines) and maximum cyclic stress amplitude–strain amplitude ($\sigma_{A,max}$-$\Delta\varepsilon/2$: solid red lines) curves for TM steel, as well as the definition of $\Delta\sigma_h(\Delta\varepsilon/2)$, $\Delta\sigma_{h,cyc}(\Delta\varepsilon/2)$, and $\Delta\sigma_A(\Delta\varepsilon/2)$, in which $\sigma^M$-$\varepsilon$ and $\sigma^M_{A,max}$-$\Delta\varepsilon/2$ are the monotonic and cyclic curves of the assumed matrix structure in the TM steel, respectively [95]. Reprinted with permission from Springer Nature: Metall. Mater. Trans. A, Copyright 2021.

In general, true cyclic hardening increment $\Delta\sigma_{h,cyc}(\Delta\varepsilon/2)$ of the TM steel (see Figure 11b) is obtained by [95]

$$\Delta\sigma_{h,cyc}(\Delta\varepsilon/2) = \sigma_{A,max}(\Delta\varepsilon/2) - \sigma^M_{A,max}(\Delta\varepsilon/2) = \Delta\sigma_{i,cyc}(\Delta\varepsilon/2) + \Delta\sigma_{t,cyc}(\Delta\varepsilon/2) + \Delta\sigma_{f,cyc}(\Delta\varepsilon/2) \tag{3}$$

where $\sigma_{A,max}(\Delta\varepsilon/2)$ and $\sigma^M_{A,max}(\Delta\varepsilon/2)$ are the maximum cyclic stress amplitudes of TM steel and its matrix, respectively. $\Delta\varepsilon$ is total strain amplitude. Additionally, $\Delta\sigma_{i,cyc}(\Delta\varepsilon/2)$, $\Delta\sigma_{t,cyc}(\Delta\varepsilon/2)$, and $\Delta\sigma_{f,cyc}(\Delta\varepsilon/2)$ represent "the long-range internal stress hardening", "the strain-induced transformation hardening", and "the forest dislocation hardening of the matrix" upon cyclic deformation, respectively, which can be formulated by

$$\Delta\sigma_{i,cyc}(\Delta\varepsilon/2) = \{(7 - 5\nu)\mu/5(1 - \nu)\} f \cdot \varepsilon_p{}^\mu \tag{4}$$

$$\Delta\sigma_{t,cyc}(\Delta\varepsilon/2) = g(\Delta f \alpha_m) \tag{5}$$

$$\Delta\sigma_{f,cyc}(\Delta\varepsilon/2) = \zeta\mu(b \cdot f \cdot \varepsilon/2r)^{1/2} \tag{6}$$

where $\nu$ is the Poisson's ratio, $\mu$ is the shear modulus, $\varepsilon_p{}^\mu$ is the eigenstrain, $f$ is the volume fraction of the second phase, $g(\Delta f \alpha_m)$ is a function of the strain-induced martensite fraction, $\zeta$ is a material constant, $b$ is the Burgers vector, and $r$ is the particle radius of the second phase. From the estimation of each cyclic hardening of Equations (4)–(6), Sugimoto et al. [95] propose that the fatigue hardening of TM steel is mainly associated with the compressive internal stress resulting from a difference in the flow stress (or plastic strain) between the matrix and the MA phase, with a small contribution from the strain-induced transformation and forest dislocation hardenings. The above theory is also applied to the fatigue hardening behavior of TPF [99] and TBF [95] steels.

The IT process enhances the smooth- and notch-fatigue limits ($\sigma_w$ and $\sigma_{wn}$) of (0.1–0.4)% C–1.5%Si–1.5%Mn TM steels and resultantly reduces the notch-sensitivity factor in a Vickers hardness range between HV350 and HV600, compared to the conventional JIS-SCM420, SCM435, and SCM440 Q&T steels (Figure 12a) [75]. However, the notch-fatigue limits are lower than those of TBF steels [100]. In addition, the notch-sensitivity factor is higher than that of 0.2%C–1.5%Si–1.5%Mn–(0–1.0)%Cr–(0–0.2)%Mo–(0–1.5)%Ni TBF steels in a hardness range below 400HV. The above-mentioned notch-sensitivity factor for fatigue limit ($q$) is defined by the following equation [101].

$$q = (K_f - 1)/(K_t - 1) \tag{7}$$

where $K_f$ and $K_t$ are the fatigue notch factor ($=\sigma_w/\sigma_{wn}$) and the elastic stress concentration coefficient ($K_t = 1.7$ in Ref. [75]), respectively.

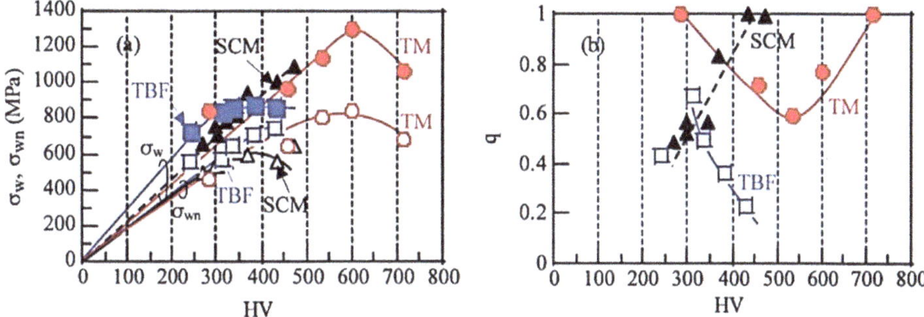

**Figure 12.** Variations in (**a**) fatigue limits ($\sigma_w$, $\sigma_{wn}$) of smooth and notched specimens and (**b**) notch-sensitivity ($q$) as a function of Vickers hardness (HV) in (0.1–0.6)%C–1.5%Si–1.5%Mn TM steels subjected to DQ-P process ($T_p$ = 250 °C, ●○), 0.2%C–1.5%Si–1.5%Mn–(0–1.0)%Cr–(0–0.2)%Mo–(0–1.5)%Ni TBF steels ($T_{IT}$ = 400 °C, ■□) and JIS-SCM420, 435 and 440 Q&T steels ($T_T$ = 200–600 °C, ▲△). This figure was reproduced from Refs. [75,100]. Refs. [75,100] are reprinted with permission from ISIJ: ISIJ Int, Copyright 2021 and from Springer nature: Metall. Mater. Trans. A, Copyright 2021, respectively.

According to Knott [102], the plastic zone size ($d_Y$) at a small fatigue crack tip can be estimated using the following equation,

$$d_Y = K^2/(3\pi YS^2) \tag{8}$$

where $K$ is the stress intensity factor defined by $K = \sigma(\pi c)^{1/2}$ ($\sigma$ is the applied stress and $c$ is the crack length) and YS is the yield stress of the material. The plastic zone size is estimated to be about 4.0 μm in 0.4%C–1.5%Si–1.5%Mn TM steel if the fatigue crack length at the first stage is 2c = 30 μm equivalent to the prior austenitic grain size and the applied stress is the maximum stress corresponding to the fatigue limit [75].

As shown in Figure 13, the plastic zone always includes some retained austenite particles in the MA phase because the inter-particle path of the MA phase is 0.5–2.0 μm, as well as a matrix structure. Based on this result, Kobayashi et al. [75] propose that the high fatigue limits and low notch-sensitivity of (0.1–0.4)%C–1.5%Si–1.5%Mn TM steels are principally associated with (i) the plastic relaxation of localized stress concentration as a result of the strain-induced transformation of 3–5 vol % metastable retained austenite and (ii) a large amount of finely dispersed MA phase along prior austenitic, packet, and block boundaries, as well as (iii) a small amount of carbide only in the coarse lath-martensite structure, which contributes to difficult fatigue crack initiation and/or propagation.

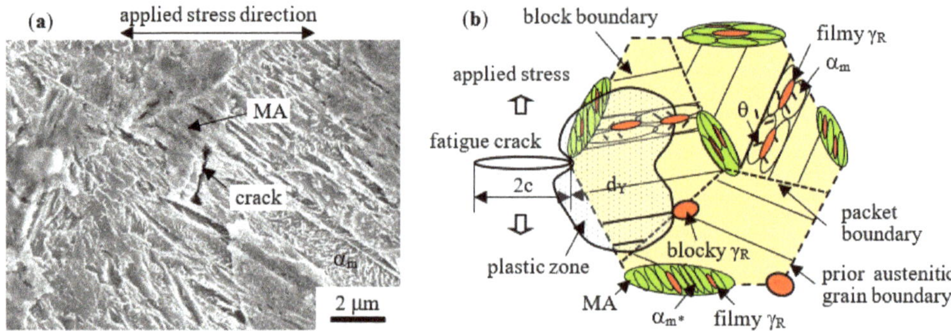


**Figure 13.** (**a**) SEM image of the initial crack formed on the notched surface of 0.4%C–1.5%Si–1.5%Mn TM steel (DQ-P, $T_p$ = 250 °C) failure at $N_f$ = 5.0 × 10$^4$ cycles [75]. (**b**) Illustration of the plastic zone size ($d_Y$) at the crack tip and distribution of MA phases in the TM steel [75]. $\alpha_m$, $\alpha_m^*$, $\gamma_R$, and θ represent coarse lath-martensite and fine lath/twin-martensite in the MA phase, retained austenite and carbide, respectively. Reprinted with permission from ISIJ: ISIJ Int, Copyright 2021.

A similar mechanism is also reported by Tomita et al. using 0.6%C–1.5%Si–0.8%Mn Q&P steel with BF/M duplex phase [103] and Huo and Gao using 0.24%C–1.75%Si–1.79%Mn–1.74%Ni–1.06%Cr–0.32%Mo–0.13%V steel subjected to the DQ-P process ($T_p$ = 150 to 550 °C) [104]. Zao et al. report that Nb addition of 0.042% in 0.21%C–1.74%Si–2.20%Mn–0.62%Cr TM steel subjected to the DQ-P process ($T_p$ = 280 °C) increases a very high cycle fatigue strength because of the cumulative effect of strengthening associated with grain refinement and precipitation strengthening mechanisms [105].

### 5.3. Delayed Fracture Strength

It is well-known that metastable retained austenite can absorb a large amount of solute hydrogen in TBF, Q&P, CFB, and D-MMn steels [57,106–110]. This results in a high delayed fracture strength in these steels because hydrogen concentration on the prior austenitic grain boundary is lowered. In 0.2%C–1.5%Si–1.5%Mn–(0–1.0)%Cr–(0–0.2)%Mo–(0–1.5%%Ni–0.05%Nb TM steels, the DQ-P process ($T_p$ = 250 and 350 °C) increases the delayed fracture strength (maximum fracture strength enduring for 5 h measured by four-point bending tests). In addition, the DQ-P process decreases hydrogen embrittlement susceptibility (HES), which is defined by the following equation in the TM steels [76,108] (Figure 14a).

$$\text{HES} = (\varepsilon_0 - \varepsilon_1)/\varepsilon_0 \times 100\% \qquad (9)$$

where $\varepsilon_0$ and $\varepsilon_1$ represent the total elongation of steel before and after hydrogen charging, respectively. The HES values are comparable with those of TBF steels with the same chemical composition and far lower than those of 0.40%C–0.16%Si–1.44%Mn Q&T steel [106–108]. The HES of the TM steels can be improved by the addition of micro-alloying elements, especially 1.0%Cr addition (corresponding to steel D of TM steel in Figure 14) [63]. In the 1.0%Cr TM steels, the metastable retained austenite plays a role in trapping the diffusible hydrogen, similar to the third-generation AHSSs (Type A) [63,76] (Figure 14b). Hojo et al. [76] propose that the high hydrogen embrittlement resistance is caused by that (i) most of the solute hydrogen is trapped in the retained austenite with high mechanical stability, and resultantly, (ii) the initiation and propagation of voids and cracks at prior austenite grain, packet, block, and lath boundaries are suppressed.

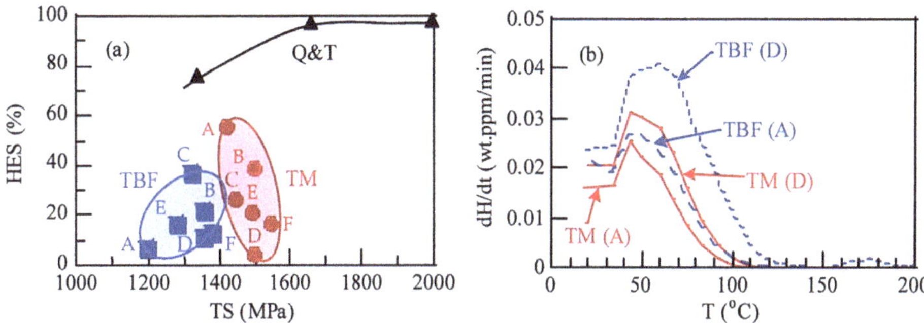

**Figure 14.** (a) Relationship between hydrogen embrittlement susceptibility (HES) and tensile strength (TS) [63] and (b) hydrogen evolution curves of 0.2%C–1.5%Si–1.5%Mn–(0–1.0)%Cr–(0–0.2)%Mo–0.05%Nb TM steels subjected to DQ-P process ($T_p$ = 250 °C, 300 °C, •) and TBF steels ($T_{IT}$ = 350 °C, ■) and 0.40%C–0.16%Si–1.44%Mn Q&T steel ($T_T$ = 200 °C to 400 °C, ▲) [63]. A: 0.20%C–1.50%Si–1.50%Mn, B: 0.20%C–1.52%Si–1.50%Mn–0.05%Nb, C: 0.21%C–1.49%Si–1.50%Mn–0.5%Cr–0.05%Nb, D: 0.20%C–1.49%Si–1.50%Mn–1.0%Cr–0.05%Nb, E: 0.18%C–1.48%Si–1.49%Mn–1.0%Cr–0.2%Mo–0.05%Nb, F: 0.21%C–1.49%Si–1.49%Mn–1.52%Ni–1.0%Cr–0.2%Mo–0.05%Nb.

Many researchers are investigating the hydrogen embrittlement of D-MMn steels [57,111,112]. Unfortunately, there is not any research on the delayed fracture strength of M-MMn steel up to now.

### 5.4. Wear Property and Overall Performance of Mechanical Properties

There are many kinds of research on the wear properties of TBF, Q&P, and CFB steels with BF/M matrix structure [97,113,114]. However, the research on the wear property of TM and M-MMn steels is hardly presented. According to De Oliveira et al. [113], the two-step Q&P process improves the wear performance compared to the one-step Q&P process ($T_{IT} > M_s$) in 0.29%C–1.35%Si–1.96%Mn–0.1%Cr–0.03%Mo–0.04%Nb steel. Feng et al. [97] show using 0.24%C–1.44%Si–1.76%Mn–0.75%Ni–1.76&Cr–0.39%Mo–0.65%Al rail steel that the DQ process (air cooling) after austenitizing enhances the wear resistance compared to the IT process above $M_s$. They propose that the strain-induced martensite transformation of retained austenite in the steel can be linked to the retardation of crack initiation and propagation, as well as martensite formed on air cooling. Wang et al. [114] show that partitioning and cryogenic treatment after the IT process ($T_{IT}$ = 360 °C and 400 °C > $M_s$) followed by slow cooling is effective to increase the wear property in 0.22%C–2.0%Mn–1.0%Si–0.8%Cr–0.8%(Mo+Ni) BF/M rail steel. This is mainly associated with a decrease in the volume fraction of retained austenite or an increase in the hardness because surface hardness is the main factor controlling the abrasive wear [115]. From the above results of TBF, Q&P, and CFB rail steels, we can expect that TM and M-MMn steels possess good wear resistance because the hardness is higher than the TBF, Q&P, and CFB steels, although the retained austenite fraction is lower.

When various mechanical properties without wear property of 0.2%C–1.5%Si–1.5Mn TM steel are compared to those of TBF steel with the same composition and DIN-22MnB5 Q&T steel, the TM steel is superior to that of Q&T steel (Figure 15) [116,117]. The mechanical properties of M-MMn steel are not ready yet. Ausforming at temperatures between $Ac_3$ and $M_s$ is expected to enhance the whole mechanical properties. However, only the effects of ausforming conditions on the tensile strength, total elongation, and impact toughness are reported up to now [118,119].

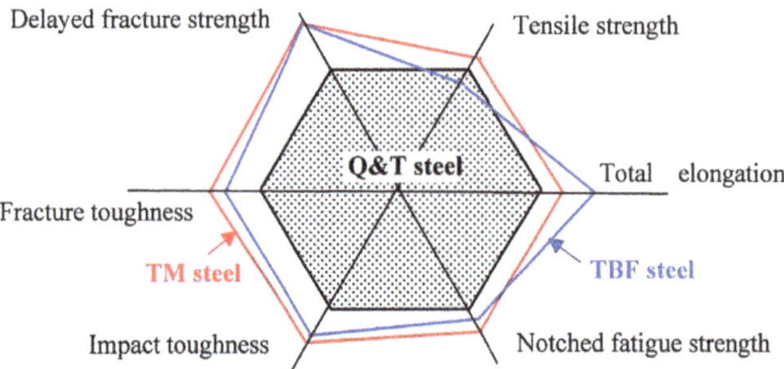



**Figure 15.** Comparison of various mechanical properties of 0.2%C–1.5%Si–1.5Mn TBF and TM steels and DIN-22MnB5 Q&T steel. This figure was modified based on Ref. [117].

## 6. Hot- and Warm-Workings to Produce Automotive AMS Parts

Hot-workings such as hot-stamping and hot-forging refine the microstructure and increase the retained austenite fraction in AMSs, in the same way as ausforming. Resultantly, the hot-workings enhance strength and other mechanical properties. In the following, the tensile and mechanical properties of hot-stamped and hot-forged AMSs are introduced.

### 6.1. Hot- and Warm-Stamping

In the conventional hot-stamping, the steel blank is completely austenitized at temperatures above $Ac_3$ before stamping, followed by die-quenching to 150 to 250 °C [120,121]. The hot-stamping process considerably reduces the spring back of the products, differing from the cold-stamping process [120]. As the process suppresses the hydrogen embrittlement due to lowering the residual stress, hot-stamping steels such as DIN-22MnB5 and 30MnB5 steels are applied to automotive center pillars with the tensile strength of 1.5 to 1.8 GPa [120,122]. The matrix structure after die-quenching is almost martensite.

Recently, warm-stamping at temperatures above $M_s$ are developed to improve productivity and reduce manufacturing costs [122–125] (Figure 16). In the warm-stamping, the blank is austenitized at temperatures above $Ac_3$, in the same way as the conventional hot-stamping process. After that, the austenitized blanks are pre-cooled to any temperatures during $Ac_3$ and $M_s$ before stamping. By this warm-stamping, the blank is strain-hardened along with the improvement of the drawability and formability. In addition, productivity can also be increased by shortening the cooling cycle time and decreasing the oxidation [122]. At present, the warm-stamping at 450 to 500 °C is applied to the B-pillar of (0.08–0.2)%C–(4.0–7.0)%Mn D-MMn steels, not M-MMn steel [125]. After the warm-stamping, the tensile strength is about 1.4 GPa with total elongation of 11.8%.

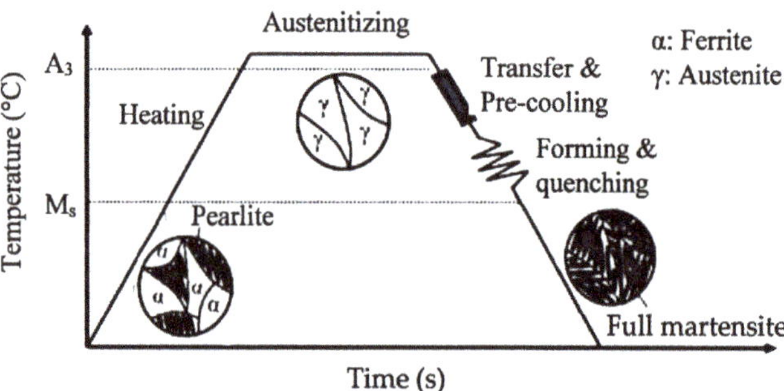



**Figure 16.** Schematic illustration of hot-stamping process at low-temperature and warm temperature [122].

As another latest advance, the combined hot-stamping and two-step Q&P process is proposed [57,126]. In 0.28%C–1.69%Si–1.10%Mn–0.99%Cr–0.029%Nb steel, in which the two-step Q&P process brings on the matrix structure containing martensite. This combined process is also applied to the TM steel subjected to the IT and DQ-P processes [127].

*6.2. Hot- and Warm-Forging*

Up to now, the conventional forging steels such as precipitation hardening ferrite/pearlite (PHFP) steels, micro-alloyed PHFP (PHFP-M) steels [128,129], bainitic steels [130,131] and Q&T martensitic steels [132,133] are applied to the automotive engine, powertrain and chassis for weight reduction. For further weight reduction of their parts, 1.5 to 2.0 GPa grade hot-forging AMSs such as TM [118,119,129] and M-MMn [78–80] steels subjected to the IT, DQ, and DQ-P processes have been recently developed, as well as TBF [80,134–136], Q&P [137], and CFB [138,139] forging steels with the BF/M matrix. In general, this prospective AMSs contains Si and/or Al higher than 1.0 mass % to suppress the carbide formation and promote a predominant formation of carbon-enriched retained austenite [118,119,140]. Micro-alloying of Mn, Cr, Ni, Mo, B, etc. is fundamental to increase the hardenability. In addition, additions of V, Ti, and/or Nb are required for refining the prior austenitic grain and precipitation hardening.

According to Kobayashi et al. [118], hot-forging at 900 °C ($>Ac_3$) with a reduction strain of 40 to 60% (one pass, strain rate: 50%/s) followed by the DQ-P process ($T_p$ = 200 to 350 °C) produces a uniform and fine microstructure in 0.4C–1.5Si–1.5Mn–0.05Nb–0.002B TM steel, without polygonal ferrite. In addition, the hot-forging increases the volume fractions of retained austenite and MA phase, with the decreased carbide fraction. Resultantly, the hot-forging and subsequent DQ-P process bring on an excellent combination of yield stress and Charpy V-notch impact value (YS = 1400 to 1561 MPa and $E_v$ of 35 to 44 J/cm$^2$) (Figure 17) [117,118]. The balance is far higher than the conventional Q&T martensitic, bainitic, PHFP, and PHFP-M steels [141–146], and it is nearly the same as those of TBF, Q&P, and CFB steels with a BF/M matrix structure, although the balance is slightly inferior to that of TBF steels with the same chemistry. According to Kobayashi et al. [118,119], the increased balance is associated with (i) refined duplex phase structure (soft coarse lath-martensite and MA phase) and (ii) increased volume fractions of retained austenite and MA phase. Note that the balance (YS×$E_v$) value of D-MMn steel is lower, although the steel possesses an extremely high TS×TEl value.

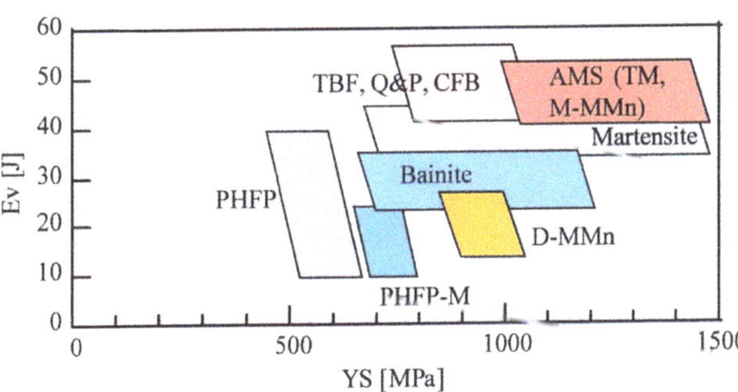


**Figure 17.** Relationship between Charpy V-notch impact value ($E_v$) and yield stress (YS) at room temperature in steel groups [77,117,141–146]. PHFP, PHFP-M, TBF, Q&P, CFB, D-MMn, AMS, TM and M-MMn are precipitation-hardening ferritic–pearlitic, micro-alloyed PHFP, TRIP-aided bainitic ferrite, quenching and partitioning, carbide-free bainitic, duplex-type medium Mn, advanced martensitic, TRIP-aided martensitic, and martensite-type medium Mn steels, respectively. This figure is reproduced based on Ref. [117].

The cooling rate just after hot-forging is a very important factor controlling the microstructure and mechanical properties of AMSs. Kobayashi et al. [119] show an important result that a low cooling rate (1.2 °C/s) to room temperature after hot-forging increases the $E_v$ and lowers the DBTT in 0.21%C–1.49%Si–1.49%Mn–1.0%Cr–0.20%Mo–1.50%Ni–0.05%Nb TM steel. In addition, they show that the improved impact toughness is mainly caused by the increased volume fraction and carbon concentration of retained austenite and the decreased carbide fraction as well as the finely dispersed MA phase. Gramlich et al. [80] show that air cooling to room temperature after hot-forging produces a mixture of martensite with a small amount of carbide and retained austenite and results in extremely high strain-hardening in 0.18%C–0.52%Si–3.98%Mn–0.54%Al–0.11%Cr–0.20%Mo–0.10%Ni–0.11%V–0.036%Nb M-MMn steel. In this case, subsequent partitioning or tempering at 250 to 350 °C slightly increases the carbide size without major changes in morphology. Resultantly, the partitioning increases the yield stress and slightly decreases the tensile strength with no change in elongation.

Hojo et al. [82] show that warm-forging at 650 °C and subsequent slow cooling (1 °C/s) increases the volume fraction and carbon concentration of the retained austenite and decreases the volume fractions of MA phase and carbide in 0.23%C–1.5%Si–1.5%Mn–1.0%Cr–0.2%Mo–0.1%Ti–0.0019%B TM steel (Figure 18). The product of shear stress and shear elongation of the TM steel after warm-forging, which is measured by small punching tests at room temperature, is slightly increased.

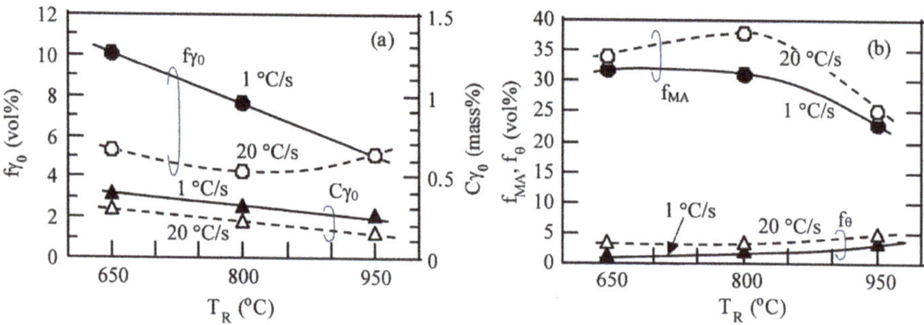

**Figure 18.** Variations in (**a**) initial volume fraction ($f_{\gamma 0}$) and initial carbon concentration ($C_{\gamma 0}$) and (**b**) volume fractions of MA phase ($f_{MA}$) and carbide ($f_\theta$) of retained austenite as a function of warm working temperature ($T_R$) in 0.23%C–1.5%Si–1.5%Mn–1.0%Cr–0.2%Mo–0.1%Ti–0.0019%B TM steel [82]. Cooling rate: 1 °C/s (solid marks) or 20 °C/s (open marks). Reduction strain: 40%. Reprinted with permission from AIST: Iron Steel Technol., Copyright 2021.

## 7. Case Hardening of AMSs

Fine particle peening (FPP) treatment after the heat-treatment (the DQ-P process, $T_p = 180$ °C) increases the rotating bending fatigue limits of smooth and notched specimens in 0.20%C–1.50%Si–1.51%Mn–1.0%Cr–0.05%Nb TM steel (Figure 19a) [147]. Vacuum carburization (VC) treatment followed by the DQ-P process and subsequent FPP treatment (VC+FPP, $T_p = 180$ °C) further increases the fatigue limits in 0.2%C–1.5%Si–1.5%Mn–1.0%Cr–0.05%Nb TM steel [148]. According to Sugimoto et al., the high fatigue limits are obtained under the conditions of carbon potential of 0.8 mass% (Figure 19b) [148] and arc-height of 0.21 mm with a gage N (Figure 19a) [147].

The fatigue limits of case-hardened TM steel are mainly controlled by the Vickers hardness (or yield stress) and compressive residual stress in the surface hardened layer, which are developed from the severe strain-hardening and strain-induced transformation of retained austenite [147,148]. As shown by line (1) in Figure 20, the smooth fatigue limit ($\sigma_w$) of 0.2%C–1.5%Si–1.5%Mn–1.0%Cr–0.05%Nb TM steel subjected to only the FPP treatment shows a linear relationship with the sum of the estimated maximum yield stress and the absolute value of maximum compressive X-ray residual stress ($\sigma_{Y,est} + |\sigma_{X\alpha,max}|$) as the following equation proposed by Matsui and Koshimune et al. [149,150],

$$\sigma_w = 0.3891 \times (\sigma_{Y,est} + |\sigma_{X\alpha,max}|), \ (2363 \text{ MPa} < \sigma_{Y,est} + |\sigma_{X\alpha,max}| < 4505 \text{ MPa}) \tag{10}$$

where $\sigma_{Y,est}$ is calculated by

$$\sigma_{Y,est} = (HV_{max}/3) \times 9.80665 \times (YS/TS) \tag{11}$$

where $HV_{max}$ is the maximum Vickers hardness and YS/TS is the yield ratio (assumed by Matsui et al. to be 0.95 in 0.22%C–0.27%Si–0.74%Mn–1.1%Cr–0.36%Mo Q&T steel gas-carburized and then shot-peened [150]).

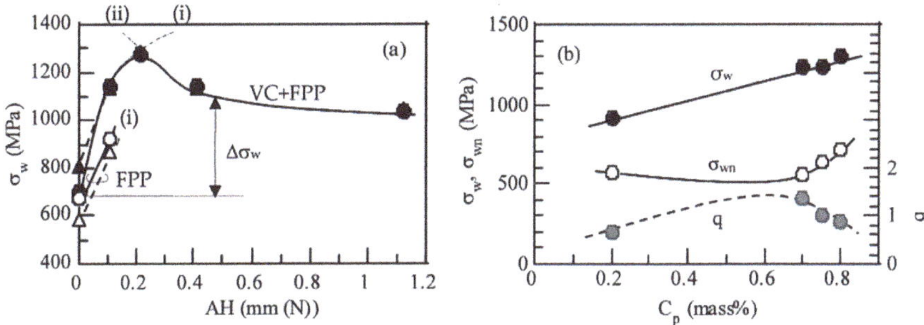

**Figure 19.** (**a**) Arc-height (AH) dependence of smooth fatigue limit ($\sigma_w$) in 0.2%C–1.5%Si–1.5%Mn–1.0%Cr–0.05%Nb TM steel (●○) and JIS-SNCM420 Q&T steel (△▲) subjected to fine particle peening (FPP) treatment after heat-treatment (the DQ-P process, $T_p$ = 180 °C) and vacuum carburization (VC) followed by the DQ-P process ($T_p$ = 180 °C) and then fine particle peening (FPP) treatment (VC+FPP) [147]. (**b**) Carbon potential ($C_p$) dependence of $\sigma_w$, notch-fatigue limit ($\sigma_{wn}$) and notch sensitivity ($q$) in the TM steel subjected to VC+FPP treatment [148]. (i) Increases in hardness and compressive residual stress, (ii) fish-eye crack fracture at a high depth. Refs. [147,148] are reprinted with permissions from Springer Nature: Metall. Mater. Trans. A, Copyright 2021 and from Taylor & Francis: Mater. Sci. Technol., Copyright 2021, respectively.

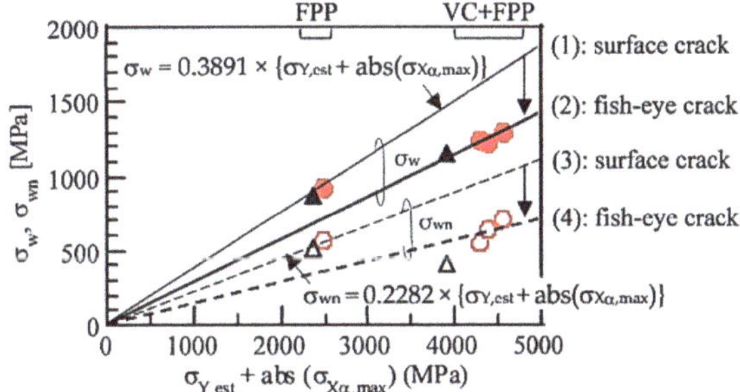

**Figure 20.** Relationships between smooth and fatigue limits ($\sigma_w$, $\sigma_{wn}$) and the sum of estimated yield stress and the absolute value of maximum compressive residual stress {$\sigma_{Y,est}$ + abs($\sigma_{X\alpha,max}$)} of smooth (solid marks) and notched (open marks) specimens in 0.2%C–1.5%Si–1.5%Mn–1.0%Cr–0.05%Nb TM (●○) and JIS-SNCM420 (▲△) steels subjected to only fine particle peening (FPP) with an arc height of 0.21 mm(N) after the heat-treatment (DQ-P process, $T_p$ = 180 °C)) and vacuum carburization under various carbon potentials ($C_P$ = 0.2 to 0.8 mass%) and subsequent FPP (VC+FPP). Lines (1) and (3) denote the fatigue limits of the smooth and notched specimens subjected to FPP treatment, respectively. Lines (2) and (4) refer to the fatigue limits of the smooth and notched specimens subjected to VC+FPP treatment [148]. Reprinted with permission from Taylor & Francis: Mater. Sci. Technol., Copyright 2021.

As shown in Figure 20, the slope in Equation (10) of notch-fatigue limit ($\sigma_{wn}$) lowers to 0.2282 in TM steel subjected to the FPP treatment (see Line (3)) [147]. On the other hand, the slops corresponding to the smooth and notch-fatigue limits of TM steel subjected to the VC+FPP treatment further decrease such as Lines (2) and (4), respectively [147,148]. The decreased slopes of Lines (2) and (4) are caused by the fish-eye crack fracture. It is noteworthy that the smooth and notch-fatigue limits of TM steel are higher than those of JIS-SNCM420 Q&T steel subjected to the same process.

Skołek et al. [151] report that DIN-35CrSiMn5-5-4 steel subjected to the DQ-P (or Q&T) process after the VC treatment (without FPP or shot-peening) possesses low surface wear resistance compared to the steel with nano-bainitic structure subjected to the IT process above $M_s$. They say that the low wear resistance of DQ-P process steel is caused by low retained austenite fraction. Kanetani et al. [152] investigate the morphology of the deformation-induced martensite formed on rolling contact fatigue in SAE4320 steel subjected to carburization and then Q&T process and show an interesting result that the deformation-induced martensite is formed preferentially within the retained austenite grains, not from the interface between tempered martensite and retained austenite.

## 8. Summary

The AMS sheets can be produced by lower heat-treatment temperature than the third-generation AHSSs (Type A) with BF/M structure matrix. As they possess the same excellent cold formability as the third-generation AHSSs (Type A), some applications to the automotive frame members can be expected in the future [153–155]. In addition, the AMSs can be applied to the automotive wire rod and bar parts such as engine, powertrain and chassis parts. Thus, the wear property, weldability, and machinability of the hot-forged AMSs also must be systematically investigated in the future, apart from toughness, fatigue strength, and delayed fracture strength described in this review. In addition, the effects of case-hardening on the mechanical properties are also required, because the engineers and scientists want to know them.

Finally, it is emphasized that the AMSs may be applied to not only automotive parts but also the forging parts of other engineering structures such as construction machinery, airplane, marine machinery, etc. Such applications will increase the fracture strength of many products and resultantly bring about a great increase in reliability. The microstructure and mechanical properties of martensitic stainless steel containing metastable retained austenite subjected to the DQ and DQ-P process should be also discussed as one kind of AMSs in the future.

**Funding:** This research received no external funding.

**Institutional Review Board Statement:** Not applicable.

**Informed Consent Statement:** Not applicable.

**Data Availability Statement:** Not applicable.

**Acknowledgments:** I thank Tomohiko Hojo from Tohoku University and Junya Kobayashi from Ibaraki University for their kind discussion.

**Conflicts of Interest:** The author declares no conflict of interest.

## References

1. Zackay, V.F.; Parker, E.R.; Fahr, D.; Bush, R. The enhancement of ductility in high-strength steels. *Trans. Am. Soc. Met.* **1967**, *60*, 252–259.
2. Tamura, I.; Maki, T.; Hato, H. On the morphology of strain-induced martensite and the transformation-induced plasticity in Fe-Ni and Fe-Cr-Ni alloys. *Trans. Iron Steel Inst. Jpn.* **1970**, *10*, 163–172. [CrossRef]
3. Matsumura, O.; Sakuma, Y.; Takechi, H. TRIP in its kinetic aspects in austempered 0.4C–1.5Si–0.8Mn steel. *Scr. Met.* **1987**, *21*, 1301–1306. [CrossRef]
4. Itami, A.; Takahashi, M.; Ushioda, K. Plastic stability of retained austenite in the cold-rolled 0.14%C-1.9%Si-1.7%Mn sheet steel. *ISIJ Int.* **1995**, *35*, 1121–1127. [CrossRef]

5.	Hiwatashi, S.; Takahashi, M.; Katayama, T.; Usuda, M. Effect of deformation-induced transformation on deep drawability—Forming mechanism of TRIP type high-strength steel sheet. *J. Jpn. Soc. Technol. Plast.* **1994**, *35*, 1109–1114.

6.	Sugimoto, K.; Kobayashi, M.; Nagasaka, A.; Hashimoto, S. Warm stretch-formability of TRIP-aided dual-phase sheet steels. *ISIJ Int.* **1995**, *35*, 1407–1414. [CrossRef]

7.	Fonstein, N. *Advanced High Strength Sheet Steels—Physical Metallurgy, Design, Processing and Properties*; Springer International Publishing: Cham, Switzerland, 2015.

8.	Speer, J.G.; Edmonds, D.V.; Rizzo, F.C.; Matlock, D.K. Partitioning of carbon from supersaturated plates of ferrite with application to steel processing and fundamentals of the bainitic transformation. *Curr. Opin. Solid State Mat. Sci.* **2004**, *8*, 219–237. [CrossRef]

9.	Rana, R.; Singh, S.B. *Automotive Steels—Design, Metallurgy, Processing and Applications*; Woodhead Publishing: Cambridge, UK, 2016.

10.	Hausmann, K.; Krizan, D.; Spiradek-Hahn, K.; Pichler, A.; Werner, E. The influence of Nb on transformation behavior and mechanical properties of TRIP-assisted bainitic–ferritic sheet steels. *Mater. Sci. Eng. A* **2013**, *588*, 142–150. [CrossRef]

11.	Grässel, O.; Krüger, L.; Frommeyer, G.; Meyer, L.W. High strength Fe-Mn-(Al, Si) TRIP/TWIP steels development—Properties—Application. *Int. J. Plast.* **2000**, *16*, 1391–1409. [CrossRef]

12.	Furukawa, T. Structure and mechanical properties of dual phase steel. *J. Jpn. Inst. Met. Mater.* **1980**, *19*, 439–446.

13.	Speich, G.R.; Demarest, V.A.; Miller, R.L. Formation of austenite during intercritical annealing of dual-phase steel. *Metall. Trans. A* **1981**, *12A*, 1419–1428. [CrossRef]

14.	Shirasawa, H.; Tanaka, Y.; Miyahara, M. Development of TS980 MPa-grade cold-rolled, high-strength steel sheet with high ductility. *Kobelco Technol. Rev.* **1987**, *1*, 13–17.

15.	Spenger, F.; Hebesberger, T.; Pichler, A.; Krempaszky, C.; Werner, E.A. AHSS steel grades: Strain hardening and damage as material design criteria. In Proceedings of the International Conference on New Developments in Advanced High Strength Sheet Steels, Orlando, FL, USA, 15–18 June 2008; pp. 39–49.

16.	Tasan, C.C.; Diehl, M.; Yan, D.; Bechtold, M.; Roters, F.; Schmmann, L.; Zheng, C.; Peranio, N.; Ponge, D.; Koyama, M.; et al. An overview of dual-phase steels: Advanced in microstructure-oriented micromechanically guided design. *Annu. Rev. Mater. Res.* **2015**, *45*, 391–431. [CrossRef]

17.	De Cooman, B.C. Structure-properties relationship in TRIP steels containing carbide-free bainite. *Curr. Opin. Solid State Mater. Sci.* **2004**, *8*, 285–303. [CrossRef]

18.	Srivastava, A.K.; Bhattacharjee, D.; Jha, G.; Gope, N.; Singh, S.S. Microstructural and mechanical characterization of C-Mn-Al-Si cold-rolled TRIP-aided steel. *Mater. Sci. Eng. A* **2007**, *445–446*, 549–557. [CrossRef]

19.	Tsuchida, N.; Okura, S.; Tanaka, T.; Toji, Y. High-speed tensile deformation behavior of 1 GPa-grade TRIP-aided multi-phase steels. *ISIJ Int.* **2018**, *58*, 978–986. [CrossRef]

20.	Sugimoto, K.; Kanda, A.; Kikuchi, R.; Hashimoto, S.; Kashima, T.; Ikeda, S. Ductility and formability of newly developed high strength low alloy TRIP-aided sheet steels with annealed martensite matrix. *ISIJ Int.* **2002**, *42*, 910–915. [CrossRef]

21.	Ding, R.; Tang, D.; Zhao, A. A novel design to enhance the amount of retained austenite and mechanical properties in low-alloyed steel. *Scr. Mater.* **2014**, *88*, 21–24. [CrossRef]

22.	Xie, Z.; Ren, Y.; Zhou, W.; Yang, J.; Shang, C.; Misra, R.D.K. Stability of retained austenite in multi-phase microstructure during austempering and its effect on the ductility of a low carbon steel. *Mater. Sci. Eng. A* **2014**, *603*, 69–75. [CrossRef]

23.	Kuziak, R.; Kawalla, R.; Waengler, S. Advanced high strength steels for automotive industry. *Arch. Civ. Mech. Eng.* **2008**, *8*, 103–117. [CrossRef]

24.	Jin, J.; Lee, Y. Effects of Al on microstructure and tensile properties of C-bearing high Mn TWIP steel. *Acta Mater.* **2012**, *60*, 1680–1688. [CrossRef]

25.	Bleck, W.; Guo, X.; Ma, Y. The TRIP effect and its application in cold formable sheet steels. *Steel Res. Int.* **2017**, *88*, 1700218. [CrossRef]

26.	De Cooman, B.C.; Estrin, Y.; Kim, S.K. Twinning-induced plasticity (TWIP) steels. *Acta Mater.* **2018**, *142*, 283–362. [CrossRef]

27.	Hamada, A.S.; Kisko, A.; Khosravifard, A.; Hassan, M.A.; Karjalainen, L.P.; Porter, D. Ductility and formability of three high-Mn TWIP steels in quasi-static and high-speed tensile and Erichsen tests. *Mater. Sci. Eng. A* **2018**, *712*, 255–265. [CrossRef]

28.	Hwang, J. Deformation behaviors of various Fe-Mn-C twinning-induced plasticity steels: Effect of stacking fault energy and chemical composition. *J. Mater. Sci.* **2020**, *50*, 1779–1795. [CrossRef]

29.	Sugimoto, K.; Sakaguchi, J.; Iida, T.; Kashima, T. Stretch-flangeability of a high-strength TRIP type bainitic steel. *ISIJ Int.* **2000**, *40*, 920–926. [CrossRef]

30.	Sugimoto, K.; Tsunezawa, M.; Hojo, T.; Ikeda, S. Ductility of 0.1-0.6C-1.5Si-1.5Mn ultra high-strength TRIP-aided sheet steels with bainitic ferrite matrix. *ISIJ Int.* **2004**, *44*, 1608–1614. [CrossRef]

31.	Sugimoto, K.; Murata, M.; Song, S. Formability of Al-Nb bearing ultrahigh-strength TRIP-aided sheet steels with bainitic ferrite and/or martensite matrix. *ISIJ Int.* **2010**, *50*, 162–168. [CrossRef]

32.	Kobayashi, J.; Song, S.; Sugimoto, K. Ultrahigh-strength TRIP-aided martensitic steels. *ISIJ Int.* **2012**, *52*, 1124–1129. [CrossRef]

33.	Weißensteiner, I.; Suppan, C.; Hebesberger, T.; Winkelhofer, F.; Clemens, H.; Maier-Kiener, V. Effect of morphological differences on the cold formability of an isothermally heat-treated advanced high-strength steel. *JOM* **2018**, *70*, 1567–1575. [CrossRef]

34.	Speer, J.G.; De Moor, E.; Findley, K.O.; Matlock, B.C.; De Cooman, B.C.; Edmonds, D.V. Analysis of microstructure evolution in quenching and partitioning automotive sheet steel. *Metall. Mater. Trans. A* **2011**, *42A*, 3591–3601. [CrossRef]

35. Wang, C.; Zhang, Y.; Cao, W.; Shi, J.; Wang, M.; Dong, H. Austenite/martensite structure and corresponding ultrahigh strength and high ductility of steels processed by Q&P techniques. *Technol. Sci.* **2012**, *55*, 1844–1851.
36. Ghosh, S.; Miettunen, I.; Somani, M.C.; Komi, J.; Porter, D. Nanolath martensite-austenite structures engineered through DQ & P processing for developing tough, ultrahigh strength steels. *Mater. Today Proc.* **2021**, *38*. in press.
37. Tan, X.; Xu, Y.; Yang, X.; Wu, D. Microstructure–properties relationship in a one-step quenched and partitioned steel. *Mater. Sci. Eng. A* **2014**, *589*, 101–111. [CrossRef]
38. Kähkönen, J.; Pierce, D.T.; Speer, J.G.; De Moor, E.; Thomas, G.A.; Coughlin, D.; Clarke, K.; Clarke, A. Quenched and partitioned CMnSi steels containing 0.3 wt.% and 0.4 wt.% carbon. *JOM* **2016**, *68*, 210–214. [CrossRef]
39. Fan, D.; Fonstein, N.; Jun, H. Effect of microstructure on tensile properties and cut-edge formability of DP, TRIP, Q & T and Q & P steels. *AIST Trans.* **2016**, *13*, 180–185.
40. Seo, E.; Cho, L.; Estrin, Y.; De Cooman, B.C. Microstructure-mechanical properties relationships for quenching and partitioning (Q & P) processed steel. *Acta Mater.* **2016**, *113*, 124–139.
41. Huyghe, P.; Dépinoy, S.; Caruso, M.; Mercier, D.; Georges, C.; Malet, L.; Godet, S. On the effect of Q & P processing on the stretch-flange-formability of 0.2C ultra-high strength steel sheets. *ISIJ Int.* **2018**, *58*, 1341–1350.
42. Caballero, F.G.; Santofimia, M.J.; Garcia-Mateo, C.; Chao, J.; Garcia de Andrés, C. Theoretical design and advanced microstructure in super high strength steels. *Mater. Des.* **2009**, *30*, 2077–2083. [CrossRef]
43. Zhang, X.; Xu, G.; Wang, X.; Embury, D.; Bouaziz, O.; Purdy, G.P.; Zurob, H.S. Mechanical behavior of carbide-free medium carbon bainitic steels. *Metall. Mater. Trans. A* **2014**, *45A*, 1352–1361. [CrossRef]
44. Rana, R.; Chen, S.; Halder, A.; Das, S. Mechanical properties of a bainitic steel producible by hot rolling. *Arch. Metall. Mater.* **2017**, *62*, 2331–2338. [CrossRef]
45. Das, S.; Sinha, S.; Lodh, A.; Chintha, A.R.; Krugla, M.; Haldar, A. Hot-rolled and continuously cooled bainitic steel with good strength-elongation combination. *Mater. Sci. Technol.* **2017**, *33*, 1026–1037. [CrossRef]
46. Garcia-Mateo, C.; Paul, G.; Somani, M.C.; Porter, D.A.; Bracke, L.; Latz, A.; De Andres, C.G.; Caballero, F.G. Transferring nanoscale bainite concept to lower contents: A prospective. *Metals* **2017**, *7*, 159. [CrossRef]
47. Tian, J.; Xu, G.; Zhou, M.; Hu, H. Refined bainite microstructure and mechanical properties of a high-strength low-carbon bainitic steel. *Steel Res. Int.* **2018**, *89*, 1700469. [CrossRef]
48. Navarro-López, A.; Hidalgo, J.; Sietsma, J.; Santofimia, M.J. Influence of the prior athermal martensite on the mechanical response of advanced bainitic steel. *Mater. Sci. Eng. A* **2018**, *735*, 343–353. [CrossRef]
49. Chen, G.; Hu, H.; Xu, G.; Tian, J.; Wan, X.; Wang, X. Optimizing microstructure and property by ausforming in a medium-carbon bainitic steel. *ISIJ Int.* **2020**, *60*, 2007–2014. [CrossRef]
50. Miller, R.I. Ultrafine-grained microstructures and mechanical properties of alloy steels. *Metall. Trans.* **1972**, *3*, 905–912. [CrossRef]
51. Furukawa, T. Dependence of strength-ductility characteristics on thermal history in low carbon, 5 wt-%Mn steels. *Mater. Sci. Technol.* **1989**, *5*, 465–470. [CrossRef]
52. Cao, W.; Wang, C.; Shi, J.; Wang, M.; Hui, W.; Dong, D. Microstructure and mechanical properties of Fe–0.2C–5Mn steel processed by ART-annealing. *Mater. Sci. Eng. A* **2011**, *528*, 6661–6666. [CrossRef]
53. Torizuka, S.; Hanamura, T. Ultra fine heterogeneous microstructure obtained by 5% Mn steel which makes it possible to achieve 10000 GPa%J in the product of strength-ductility-toughness balance. *Bull. Iron Steel Inst. Jpn.* **2012**, *17*, 852–857.
54. Hu, B.; He, B.; Cheng, G.; Yen, H.; Huang, M.; Luo, H. Super-high-strength and formable medium Mn steel manufactured by warm rolling process. *Acta Mater.* **2019**, *174*, 131–141. [CrossRef]
55. Sugimoto, K.; Tanino, H.; Kobayashi, J. Warm stretch-formability of 0.2%C-1.5%Si-(1.5–5.0)%Mn TRIP-aided steels. *Arch. Mater. Sci. Eng.* **2016**, *80*, 5–15. [CrossRef]
56. Ma, Y. Medium-manganese steels processed by austenite-reverted-transformation annealing for automotive applications. *Mater. Sci. Technol.* **2017**, *33*, 1713–1727. [CrossRef]
57. Cho, L.; Kong, Y.; Speer, J.G.; Findley, K.O. Hydrogen embrittlement of medium Mn steels. *Metals* **2021**, *11*, 358. [CrossRef]
58. Wang, X.; Zhong, N.; Rong, Y.; Hsu, T. Novel ultrahigh strength nanolath martensitic steel by quenching-partitioning-tempering process. *J. Mater. Res.* **2009**, *24*, 260–267. [CrossRef]
59. Sugimoto, K.; Kobayashi, J. Newly developed TRIP-aided martensitic steels. In Proceedings of the Materials Science and Technology and Exhibition 2010 (MS & T 10), Houston, TX, USA, 17–21 October 2010; pp. 1639–1649.
60. Kobayashi, J.; Pham, D.V.; Sugimoto, K. Stretch-flangeability of 1.5GPa grade TRIP-aided martensitic cold rolled sheet steels. *Steel Res. Int. Spec. Ed. ICTP 2011* **2011**, 598–603.
61. Sugimoto, K.; Kobayashi, J.; Pham, D.V. Advanced ultrahigh-strength TRIP-aided martensitic sheet steels for automotive applications. In Proceedings of the New Developments in Advanced High Strength Sheet Steels (AIST 2013), Vail, CO, USA, 23–27 June 2013; pp. 175–184.
62. Pham, D.V.; Kobayashi, J.; Sugimoto, K. Effects of microalloying on stretch-flangeability of TRIP-aided martensitic sheet steel. *ISIJ Int.* **2014**, *54*, 1943–1951. [CrossRef]
63. Sugimoto, K.; Srivastava, A.K. Microstructure and mechanical properties of a TRIP-aided martensitic steel. *Metallogr. Microstruct. Anal.* **2015**, *4*, 344–354. [CrossRef]
64. Hanamura, T.; Torizuka, S.; Tamura, S.; Enokida, S.; Takechi, H. Effect of austenite grain size on transformation behavior, microstructure and mechanical properties of 0.1C-5Mn martensitic steel. *ISIJ Int.* **2013**, *53*, 2218–2225. [CrossRef]

65. Sugimoto, K.; Tanino, H.; Kobayashi, J. Impact toughness of medium-Mn transformation-induced plasticity-aided steels. *Steel Res. Int.* **2015**, *86*, 1151–1160. [CrossRef]
66. Sugimoto, K.; Tanino, H. Cold formability of martensite type medium Mn steels. Unpublished work.
67. He, B.; Liu, L.; Huang, M.X. Room-temperature quenching and partitioning steel. *Metall. Mater. Trans. A* **2018**, *49*, 3167–3172. [CrossRef]
68. He, B.; Wang, M.; Huang, M.X. Improving tensile properties of room-temperature quenching and partitioning steel by dislocation engineering. *Metall. Mater. Trans. A* **2019**, *50A*, 4021–4026. [CrossRef]
69. Du, P.; Chen, P.; Misra, D.K.; Wu, D.; Yi, H. Transformation-induced ductility of reverse austenite evolved by low-temperature tempering of martensite. *Metals* **2020**, *10*, 1343. [CrossRef]
70. Gao, G.; Zhang, H.; Tan, Z.; Liu, W.; Bai, B. A carbide-free bainite/martensite/austenite triplex steel with enhanced mechanical properties treated by a novel quenching–partitioning–tempering process. *Mater. Sci. Eng. A* **2013**, *559*, 165–169. [CrossRef]
71. Gao, G.; Zhang, H.; Gui, X.; Luo, P.; Tan, Z.; Bai, B. Enhanced ductility and toughness in an ultrahigh-strength Mn-Si-Cr-C steel: The great potential of ultrafine filmy retained austenite. *Acta Mater.* **2014**, *76*, 425–433. [CrossRef]
72. Gao, G.; An, B.; Zhang, H.; Guo, H.; Gui, X.; Bai, B. Concurrent enhancement of ductility and toughness in an ultrahigh strength lean alloy steel treated by bainite-based quenching-partitioning-tempering process. *Mater. Sci. Eng. A* **2017**, *702*, 104–122. [CrossRef]
73. Somani, M.C.; Porter, D.A.; Karjalainen, L.P.; Suikkanen, P.P.; Misra, R.D.K. Process design for tough ductile martensitic steels through direct quenching and partitioning. *Mater. Today Proc.* **2015**, *2S*, S631–S634. [CrossRef]
74. Kobayashi, J.; Ina, D.; Nakajima, Y.; Sugimoto, K. Effects of microalloying on the impact toughness of ultrahigh-strength TRIP-aided martensitic steel. *Metall. Mater. Trans. A* **2013**, *44A*, 5006–5017. [CrossRef]
75. Kobayashi, J.; Yoshikawa, N.; Sugimoto, K. Notch-fatigue strength of advanced TRIP-aided martensitic steels. *ISIJ Int.* **2013**, *53*, 1479–1486. [CrossRef]
76. Hojo, T.; Kobayashi, J.; Sugimoto, K.; Nagasaka, A.; Akiyama, E. Effects of alloying elements addition on delayed fracture properties of ultra high-strength TRIP-aided martensitic steels. *Metals* **2020**, *10*, 6. [CrossRef]
77. Gramlich, A.; Emmrich, R.; Bleck, W. Annealing of 4 wt.-% manganese steels for automotive forging application. In Proceedings of the 4th International Conference on Medium and High Mn Steels, Aachen, Germany, 31 March–3 April 2019; pp. 283–286.
78. Gramlich, A.; Emmrich, R.; Bleck, W. Austenite reversion tempering-annealing of 4 wt.% manganese steels for automotive forging application. *Metals* **2019**, *9*, 575. [CrossRef]
79. Gramlich, A.; Schäfers, H.; Krupp, U. Influence of alloying elements on the dynamic recrystallization of 4 wt.–% medium manganese steels. *Materials* **2020**, *13*, 5178. [CrossRef] [PubMed]
80. Gramlich, A.; Middleton, A.; Schmidt, R.; Krupp, U. On the influence of vanadium on air-hardening medium manganese steels for sustainable forging products. *Steel Res. Int.* **2021**, 2000592. [CrossRef]
81. Zhang, M.; Wang, Y.; Zheng, C.; Zhang, F.; Wang, T. Austenite deformation behavior and the effect of ausforming process on martensite starting temperature and ausformed martensite microstructure in medium-carbon Si-Al-rich alloy steel. *Mater. Sci. Eng. A* **2014**, *596*, 9–14. [CrossRef]
82. Hojo, T.; Kobayashi, J.; Kochi, T.; Sugimoto, K. Effects of thermomechanical processing on microstructure and shear properties of 22SiMnCrMoB TRIP-aided martensitic steel. *Iron Steel Technol.* **2015**, *12*, 102–110.
83. Hojo, T.; Kochi, T.; Sugimoto, K. Effects of warm working on microstructural and shear deformation properties of TRIP-aided martensitic steel. *Mater. Sci. Forum* **2016**, *879*, 2312–2317. [CrossRef]
84. Sugimoto, K.; Itoh, M.; Hojo, T.; Hashimoto, S.; Ikeda, S.; Arai, G. Microstructure and mechanical properties of ausformed ultrahigh-strength TRIP-aided steels. *Mater. Sci. Forum* **2007**, *539–543*, 4309–4314. [CrossRef]
85. Zhao, L.; Qian, L.; Zhou, Q.; Li, D.; Wang, T.; Jia, Z.; Zhang, F.; Meng, J. The combining effects of ausforming and below-Ms or above-Ms austempering on the transformation kinetics, microstructure and mechanical properties of low-carbon bainitic steel. *Mater. Des.* **2019**, *183*, 108123. [CrossRef]
86. Koistinen, D.P.; Marburger, R.E. A general equation prescribing the extent of the austenite-martensite transformation in pure iron-carbon alloys and plain carbon steels. *Acta Metall.* **1959**, *7*, 59–60. [CrossRef]
87. Deng, B.; Yang, D.; Wang, G.; Hou, Z.; Yi, H. Effects of austenitizing temperature on tensile and impact properties of a martensitic stainless steel containing metastable retained austenite. *Materials* **2021**, *14*, 1000. [CrossRef] [PubMed]
88. Sakaki, T.; Sugimoto, K.; Fukuzato, T. Role of internal stress for continuous yielding of dual-phase steels. *Acta Metall.* **1983**, *31*, 1737–1746. [CrossRef]
89. Kobayashi, J.; Tonegawa, H.; Sugimoto, K. Cold formability of 22SiMnCrB TRIP-aided martensitic sheet steel. *Proc. Eng.* **2014**, *81*, 1336–1341. [CrossRef]
90. Sugimoto, K.; Hojo, T.; Srivastava, A.K. An overview of fatigue strength of case-hardening TRIP-aided martensitic steels. *Metals* **2018**, *8*, 355. [CrossRef]
91. Kobayashi, J.; Ina, D.; Sugimoto, K. Fracture toughness of a 1.5 GPa grade C-Si-Mn-Cr-Nb TRIP-aided martensitic steel. In Proceedings of the MS & T 2012, TMS, Pittsburgh, PA, USA, 7–11 October 2012; pp. 937–944.
92. Kobayashi, J.; Ina, D.; Futamura, A.; Sugimoto, K. Fracture toughness of an Advanced ultrahigh-strength TRIP-aided steel. *ISIJ Int.* **2014**, *54*, 955–962. [CrossRef]

93. Wu, R.; Li, W.; Zhou, S.; Zhoung, Y.; Wang, L.; Jin, X. Effect of retained austenite on the fracture toughness of quenching and partitioning (Q & P)-treated sheet steels. *Metall. Mater. Trans. A* **2014**, *45A*, 1892–1902.
94. Bhadeshia, H.K.D.H. Some phase transformations in steels. *Mater. Sci. Technol.* **1999**, *15*, 22–29. [CrossRef]
95. Sugimoto, K.; Hojo, T. Fatigue hardening behavior of 1.5 GPa grade transformation-induced plasticity-aided martensitic steel. *Metall. Mater. Trans. A* **2016**, *47A*, 5272–5278. [CrossRef]
96. Sugimoto, K.; Kobayashi, M.; Inoue, K.; Masuda, Y. Fatigue-hardening Behavior of TRIP-aided Bainitic Steels. *Tetsu Hagane* **1999**, *85*, 856–862. [CrossRef]
97. Feng, X.; Zhang, F.; Kang, J.; Yang, Z.; Long, X. Sliding wear and low cycle fatigue properties of new carbide free bainitic rail steel. *Mater. Sci. Technol.* **2014**, *30*, 1410–1418. [CrossRef]
98. Glage, A.; Weidner, A.; Biermann, H. Effect of austenite stability on the low cycle fatigue behavior and microstructure of high alloyed metastable austenitic cast TRIP-steels. *Proc. Eng.* **2010**, *2*, 2085–2094. [CrossRef]
99. Sugimoto, K.; Kobayashi, M.; Yasuki, S. Cyclic deformation behavior of a transformation-induced plasticity–aided dual-phase steel. *Metall. Mater. Tran. A* **1997**, *28A*, 2637–2643. [CrossRef]
100. Yoshikawa, N.; Kobayashi, J.; Sugimoto, K. Notch-fatigue properties of advanced TRIP-aided bainitic ferrite steels. *Metall. Mater. Trans. A* **2012**, *43A*, 4129–4136. [CrossRef]
101. Dieter, G.E. *Mechanical Metallurgy*; SI Metric Edition; McGraw-Hill Book Company: London, UK, 1988; p. 403.
102. Knott, J.F. *Fundamentals of Fracture Mechanics*; Butterworth Group: Baifukan, Tokyo, 1977; p. 127.
103. Tomita, Y.; Kijima, F.; Morioka, K. Modified austempering effect on bending fatigue properties of Fe-0.6C-1.5Si-0.8Mn steel. *Zeitschrift Metallkunde* **2000**, *91*, 43–46.
104. Huo, C.; Gao, H. Strain-induced martensitic transformation in fatigue crack tip zone for a high strength steel. *Mater. Charact.* **2005**, *55*, 12–18. [CrossRef]
105. Zhao, P.; Cheng, C.; Gao, G.; Hui, W.; Misra, R.D.K.; Bai, B.; Weng, Y. The potential significance of microalloying with niobium in governing very high cycle fatigue behavior of bainite/martensite multiphase steels. *Mater. Sci. Eng. A* **2016**, *650*, 438–444. [CrossRef]
106. Hojo, T.; Sugimoto, K.; Mukai, Y.; Akamizu, H.; Ikeda, S. Hydrogen embrittlement properties of ultra high-strength low alloy TRIP-aided steels with bainitic ferrite matrix. *Tetsu Hagane* **2006**, *92*, 83–89. [CrossRef]
107. Hojo, T.; Sugimoto, K.; Mukai, Y.; Ikeda, S. Delayed fracture properties of aluminum bearing ultrahigh strength low alloy TRIP-aided steels. *ISIJ Int.* **2008**, *48*, 824–829. [CrossRef]
108. Hojo, T.; Kobayashi, J.; Sugimoto, K. Hydrogen embrittlement resistances of alloying elements adding ultra high strength low alloy TRIP-aided steels. In Proceedings of the MS & T 2012, TMS, Pittsburgh, PA, USA, 7–11 October 2012; pp. 1186–1193.
109. Zhu, X.; Li, W.; Zhao, H.; Wang, L.; Jin, X. Hydrogen trapping sites and hydrogen-induced cracking in high strength quenching & partitioning (Q & P) treated steel. *Int. J. Hydrog. Energy* **2014**, *39*, 3031–3040.
110. Hojo, T.; Kumai, B.; Koyama, M.; Akiyama, E.; Waki, H.; Saito, H.; Shiro, H.; Yasuda, R.; Shobu, T.; Nagasaka, A. Hydrogen embrittlement resistance of pre-strained ultra-high-strength low alloy TRIP-aided steel. *Int. J. Fract.* **2020**, *224*, 253–260. [CrossRef]
111. Han, J.; Nam, J.; Lee, Y. The mechanism of hydrogen embrittlement in intercritically annealed medium Mn TRIP steel. *Acta Mater.* **2016**, *113*, 1–10. [CrossRef]
112. Zhang, Y.; Hui, W.; Zhao, X.; Wang, C.; Cao, W.; Dong, H. Effect of reverted austenite fraction on hydrogen embrittlement of TRIP-aided medium Mn steel. *Eng. Fail. Anal.* **2019**, *97*, 605–616. [CrossRef]
113. De Oliveira, P.G.B.; Aureliano, R.T.J.; Casteletti, L.C.; Filho, A.I.; Neto, A.L.; Totten, G.E. Adhesive wear resistance of low temperature austempered and quenched and partitioned niobium alloyed steels. In Proceedings of the 30th ASM Heat Treating Society Conference, Detroit, MI, USA, 15–17 October 2019; pp. 193–199.
114. Wang, K.; Tan, Z.; Gu, K.; Gao, B.; Gao, G.; Misra, R.D.K.; Bai, B. Effect of deep cryogenic treatment on structure-property relationship in an ultrahigh strength Mn-Si-Cr bainite/martensite multiphase rail steel. *Mater. Sci. Eng. A* **2017**, *684*, 559–566. [CrossRef]
115. Haiko, O.; Miettunen, I.; Porter, D.; Ojala, N.; Ratia, V.; Heino, V.; Kemppainen, A. Effect of finish rolling and quench stop temperatures on impact-abrasive wear resistance of 0.35 % carbon direct-quenched steel. *Tribol. Fin. J. Tribol.* **2017**, *35*, 5–21.
116. Sugimoto, K. Fracture strength and toughness of ultrahigh strength TRIP aided steels. *Mater. Sci. Technol.* **2009**, *25*, 1108–1117. [CrossRef]
117. Sugimoto, K.; Hojo, T.; Srivastava, A.K. Low and medium carbon advanced high-strength forging steels for automotive applications. *Metals* **2019**, *9*, 1263. [CrossRef]
118. Kobayashi, J.; Sugimoto, K.; Arai, G. Effects of hot forging process on combination of strength and toughness in ultrahigh-strength TRIP-aided martensitic steels. *Adv. Mater. Res.* **2012**, *409*, 696–701. [CrossRef]
119. Kobayashi, J.; Nakajima, Y.; Sugimoto, K. Effects of cooling rate on impact toughness of an ultrahigh-strength TRIP-aided martensitic steel. *Adv. Mater. Res.* **2014**, *922*, 366–371. [CrossRef]
120. Senuma, T.; Takemoto, Y. Formability and productivity improving technology of hot stamping. *Tetsu Hagane* **2014**, *100*, 1481–1489. [CrossRef]
121. Maeno, T.; Mori, K.; Fujimoto, M. Improvements in productivity and formability by water and die quenching in hot stamping of ultra-high strength steel parts. *CIRP Ann.* **2015**, *64*, 281–284. [CrossRef]

122. Tong, C.; Rong, Q.; Yarley, V.; Li, X.; Luo, J.; Zhu, G.; Shi, Z. New developments and future trends in low-temperature hot stamping technologies: A review. *Metals* **2020**, *10*, 1652. [CrossRef]

123. Mori, K.; Maki, S.; Tanaka, Y. Warm and hot stamping of ultra high tensile strength steel sheets using resistance heating. *CIRP Ann.* **2005**, *54*, 209–212. [CrossRef]

124. Pan, H.; Cai, M.; Ding, H.; Huang, H.; Zhu, B.; Wang, Y.; Zhang, Y. Microstructure evolution and enhanced performance of a novel Nb-Mo microalloyed medium Mn alloy fabricated by low-temperature rolling and warm stamping. *Mater. Des.* **2017**, *134*, 352–360. [CrossRef]

125. Chang, Y.; Wang, C.; Zhao, K.; Dong, H.; Yan, J. An introduction to medium-Mn steel: Metallurgy, mechanical properties and warm stamping process. *Mater. Des.* **2016**, *94*, 424–432. [CrossRef]

126. Zhu, B.; Zhu, J.; Wang, Y.; Rolfe, B.; Wang, Z.; Zhang, Y. Combined hot stamping and Q & P processing with a hot air partitioning device. *J. Mater. Process. Technol.* **2018**, *262*, 392–402.

127. Mohrbacher, H. Martensitic automotive steel sheet—Fundamentals and Metallurgical optimization strategies. *Adv. Mater. Res.* **2015**, *1063*, 130–142. [CrossRef]

128. Yoshida, H.; Isogawa, S.; Ishikawa, T. Basic properties of ferrite-pearlitic microalloyed steel—Development of microalloyed steel for controlled forging. *J. Jpn. Soc. Technol. Plast.* **2001**, *41*, 569–573.

129. Clarke, P.; Green, M.; Dolman, R. Development of high strength microalloyed steel for powertrain applications. In Proceedings of the 4th International Conference on Steels for Cars and Trucks (SCT2014), Braunschweig, Germany, 15–19 June 2014; pp. 442–449.

130. Bleck, W.; Prahl, U.; Hirt, G.; Bambach, M. Designing new forging steels by ICMPE. In *Advances in Production Technology*; Chapter 7; Brecher, C., Ed.; Springer: Cham, Switzerland; Heidelberg, Germany, 2015; pp. 85–98.

131. Maminska, K.; Roth, A.; d'Eramo, E.; Marchal, F.; Galtier, A.; Sourmail, T. A new bainitic forging steel for surface induction hardened components. In Proceedings of the 5th International Conference on Steels for Cars and Trucks (SCT2017), Amsterdam, The Netherland, 18–22 June 2017; p. 228.

132. Yoshida, H.; Isogawa, S.; Ishikawa, T. Basic property of martensitic microalloyed steel—Development of microalloyed steel for forging using ausforming. *J. Jpn. Soc. Technol. Plast.* **2000**, *41*, 379–383.

133. Riedner, S.; Van Soest, F.; Kunow, S. B-alloyed quench and temper steels—An option to classic CrMo and CrNiMo quench and temper steels. In Proceedings of the 3rd International Conference on Steels for Cars and Trucks (SCT 2011), Salzburg, Austria, 5–9 June 2011; pp. 219–225.

134. Sugimoto, K.; Sato, S.; Arai, G. Hot forging of ultrahigh-strength TRIP-aided steel. *Mater. Sci. Forum* **2010**, *638–642*, 3074–3079. [CrossRef]

135. Sugimoto, K.; Sato, S.; Kobayashi, J.; Srivastava, A.K. Effects of Cr and Mo on mechanical properties of hot-forged medium carbon TRIP-aided bainitic ferrite steels. *Metals* **2019**, *9*, 1066. [CrossRef]

136. Sugimoto, K.; Sato, S.; Arai, G. The effects of hot-forging on mechanical properties of ultrahigh-strength TRIP-aided steels. In Proceedings of the International Steel Technologies Symposium (Taiwan 2008), Kaohsiung, Taiwan, 2–5 November 2008; p. D12.

137. Mašek, B.; Jirková, H. New material concept for forging. In Proceedings of the International Conference on Advances in Civil Structural Mechanical Engineering (ACSME 2014), Bangkok, Thailand, 26–27 August 2014; pp. 92–96.

138. Garcia-Mateo, C.; Morales-Rivas, L.; Caballero, F.G.; Milbourn, D.; Sourmail, T. Vanadium effect of medium carbon forging steel. *Metals* **2016**, *6*, 130. [CrossRef]

139. Zhang, F.; Yang, Y.; Shan, Q.; Li, Z.; Bi, Z.; Zhou, R. Microstructure evolution and mechanical properties of 0.4C-Si-Mn-Cr steel during high temperature deformation. *Materials* **2020**, *13*, 172. [CrossRef]

140. Sugimoto, K.; Kobayashi, J.; Nakajima, Y.; Kochi, T. The effects of cooling rate on retained austenite characteristics of a 0.2C-1.5Si-1.5Mn-1.0Cr-0.05Nb TRIP-aided martensitic steel. *Mater. Sci. Forum* **2014**, *783–786*, 1015–1020. [CrossRef]

141. Langeborg, R.; Sandberg, O.; Roberts, W. *Fundamentals of Microalloying Forging Steels*; Krauss, G., Banerji, S.K., Eds.; TMS: Warrendale, PA, USA, 1987; pp. 39–54.

142. Keul, C.; Wirths, V.; Bleck, W. New bainitic steel for forgings. *Arch. Civ. Mech. Eng.* **2012**, *12*, 119–125. [CrossRef]

143. Raedt, H.W.; Spekenheuer, U.; Vollrath, K. New forged steels—Energy-efficient solutions for stronger parts. *ATZ Worldw.* **2012**, *114*, 5–9. [CrossRef]

144. Buchmayr, B. Critical assessment 22: Bainitic forging steels. *Mater. Sci. Technol.* **2016**, *32*, 517–522. [CrossRef]

145. Bleck, W.; Bambach, M.; Wirths, V.; Stieben, A. Microalloyed engineering steels with improved performance—An Overview. *Heat Treat. Mater. J.* **2017**, *72*, 355–364. [CrossRef]

146. Sugimoto, K.; Kobayashi, J.; Arai, G. Development of low alloy TRIP-aided steel for hot-forging parts with excellent toughness. *Steel Res. Int. Spec. Ed. Met. Form.* **2010**, *81*, 254–257.

147. Sugimoto, K.; Hojo, T.; Mizuno, Y. Effects of fine particle peening conditions on the rotational bending fatigue strength of a vacuum-carburized transformation-induced plasticity-aided martensitic Steel. *Metall. Mater. Trans. A* **2018**, *49A*, 1552–1560. [CrossRef]

148. Sugimoto, K.; Hojo, T.; Mizuno, Y. Fatigue strength of a vacuum-carburized TRIP-aided martensitic steel. *Mater. Sci. Technol.* **2018**, *34*, 743–750. [CrossRef]

149. Matsui, K.; Koshimune, M.; Takahashi, K.; Ando, K. Influence of shot peening method on rotating bending fatigue limit for high strength steel. *Trans. Jpn. Soc. Spring Eng.* **2010**, *55*, 7–12. [CrossRef]

150. Koshimune, M.; Matsui, K.; Takahashi, K.; Nakano, W.; Ando, K. Influence of hardness and residual stress on fatigue limit for high strength steel. *Trans. Jpn. Soc. Spring Eng.* **2009**, *54*, 19–26. [CrossRef]
151. Skołek, E.; Wasiak, K.; Świątnicki, W.A. Structure and properties of the carburized surface layer on 35CrSiMn5-5-4 steel after nanostructurization treatment. *Mater. Technol.* **2015**, *49*, 933–939.
152. Kanetani, K.; Moronaga, T.; Hara, T.; Ushioda, K. Deformation-induced martensitic transformation behavior of retained austenite during rolling contact in carburized SAE4320 steel. *Tetsu-to-Hagane* **2020**, *106*, 953–960. [CrossRef]
153. Billur, E.; Altan, T. Three generations of advanced high-strength steels for automotive applications, Part III. *Stamp. J.* **2014**, March/April. 12–13.
154. Keeler, S.; Kimchi, M.; Mconey, P.J. *Advanced High-Strength Steels Application Guidelines Version 6.0.*; WorldAutoSteel: Brussels, Belgium, 2017.
155. Krizan, D.; Steinedei, K.; Kaar, S.; Hebesberger, T. Development of third generation advanced high strength steels for automotive applications. In Proceedings of the 19th International Scientific Conference Transfer 2018, Bansko, Bulgaria, 14–16 December 2018; pp. 1–15.

 *metals* 

*Review*

# Low and Medium Carbon Advanced High-Strength Forging Steels for Automotive Applications

**Koh-ichi Sugimoto [1,*], Tomohiko Hojo [2] and Ashok Kumar Srivastava [3]**

[1]  Department of Mechanical Systems Engineering, School of Science and Technology, Shinshu University, Nagano 380-8553, Japan

[2]  Institute for Materials Research, Tohoku University, Sendai 980-8577, Japan; hojo@imr.tohoku.ac.jp

[3]  Department of Metallurgical Engineering, School of Engineering, OP Jindal University, Raigarh 496109, India; ashok.srivastava@opju.ac.in

*  Correspondence: sugimot@shinshu-u.ac.jp; Tel.: +81-90-9667-4482

Received: 28 October 2019; Accepted: 21 November 2019; Published: 26 November 2019

**Abstract:** This paper presents the microstructural and mechanical properties of low and medium carbon advanced high-strength forging steels developed based on the third generation advanced high-strength sheet steels, in conjunction with those of conventional high-strength forging steels. Hot-forging followed by an isothermal transformation process considerably improved the mechanical properties of the forging steels. The improvement mechanisms of the mechanical properties were summarized by relating to the matrix structure, the strain-induced transformation of metastable retained austenite, and/or a mixture of martensite and austenite.

**Keywords:** advanced high-strength forging steel; hot-forging; microstructure; retained austenite characteristics; mechanical properties; applications

## 1. Introduction

Conventional high-strength forging steels (CFSs) are necessary for the production of automotive powertrain and chassis parts. Most of the CFSs are generally subjected to the heat-treatment of quenching and tempering (Q&T) after cold- or hot-forging to increase the yield stress, impact toughness, fatigue strength, etc. Because the Q&T treatment is expensive, modified V-microalloying precipitation hardening ferritic/pearlitic (PHFP-M) steels and bainitic steels subjected to hot-forging and then controlling cooling were developed to eliminate the additional Q&T treatment [1–4]. Although both the steels are applied to the automotive powertrain and chassis forging parts, their mechanical properties are inferior to those of Q&T steels. Recently, non-heat-treated forging steels with further high mechanical properties are required to achieve the weight reduction and size-down of the automotive parts [3,4].

In the past decades, the following first, second, and third generation advanced high-strength steel (AHSS) sheets shown in Figure 1 have been developed for the weight reduction and the improvement of crush safety of the automotive body.

(I)   First generation AHSS: ferrite–martensite dual-phase steels, transformation-induced plasticity (TRIP)-aided steels [5] with polygonal ferrite matrix structure, and complex phase steels [6–8].

(II)  Second generation AHSS: twinning-induced plasticity (TWIP) high Mn steels [7–10].

(III) Third generation AHSS: TRIP-aided steels with bainitic ferrite, bainitic ferrite–martensite, and martensite matrix structures (TBF, TBM, and TM steels, respectively) [11–16], quench and partitioning (Q&P) steels [7,8,17–19], nanostructured bainitic (Nano-B) steels (or carbide free bainitic steels) [3,4,7,8,20–23], dual-phase type and martensite type medium Mn (M-Mn) steels [7,8,24–26], and maraging-TRIP steels [7,27].

The third generation AHSS sheets are the most needed steel sheet grade in the future by the automotive industry. In the third generation AHSS sheets, a product or combination of tensile strength and total elongation higher than 30 GPa% is made the target (Figure 1). Low and medium carbon TBF [8,11,12], TBM [11–15], and TM steels [12–16] are produced by a similar heat-treatment process to low and medium carbon Q&P, Nano-B, and martensite type M-Mn steels, except Q&P steel subjected to a two-step Q&P process. For the TBF, TBM, and TM steels, isothermal transformation (IT) processes at temperature ($T_{IT}$) above the martensite-start temperature ($M_S$) [11,12], between $M_S$ and the martensite-finish temperature ($M_f$) [11–15] and lower than $M_f$ are conducted after austenitizing or annealing [12–16], respectively. Recently, low and medium carbon TBF [28–30] and Q&P steels [31] were applied to the automotive body and seats because their steels possess a superior combination of tensile strength and cold formability.

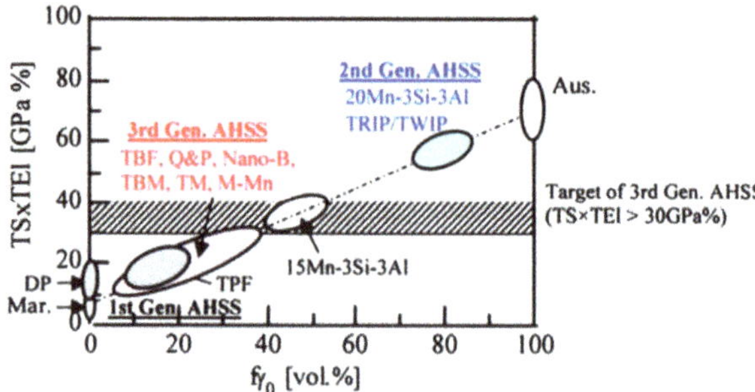


**Figure 1.** Relationship between the product of tensile strength and total elongation (TS × TEl) and initial volume fraction of austenite or retained austenite ($f\gamma_0$) in the first, second, and third generation advanced high-strength steel (AHSS) sheets. Mar.: conventional martensitic steel, DP: ferrite–martensite dual-phase steel, TPF, TBF, TBM, and TM: transformation-induced plasticity (TRIP)-aided steels with polygonal ferrite, bainitic ferrite, bainitic ferrite–martensite, and martensite matrix, respectively, Q&P: one-step and two-step quenching and partitioning steel, Nano-B: nanostructured bainitic steel, M-Mn: dual-phase type and martensite type medium Mn steel, TWIP: twinning-induced plasticity steel, Aus.: austenitic steel. This figure was modified based on Ref. [24].

The advanced high-strength forging steels (AFSs) which were developed based on the third generation AHSSs are also very attractive to apply to the automotive hot-forging parts because they bring on the especially high yield stress and tensile strength, large elongations and high impact toughness, high fatigue strength and high delayed fracture strength [32–37]. It is well known that the improved mechanical properties are associated with refined microstructure and improved retained austenite characteristics [32–36].

This paper introduces the microstructural and mechanical properties of the prospective low and medium carbon AFSs with bainitic ferrite and/or martensite matrix structure and metastable retained austenite, along with the low and medium carbon CFSs. In addition, the improvement mechanisms of the mechanical properties are detailed by relating to the matrix structure, the strain-induced transformation of metastable retained austenite and/or a mixture of martensite and austenite (MA). Unfortunately, this paper does not include various properties of dual-phase type M-Mn [24,25] and maraging-TRIP steels [27], because heat-treatment process and chemical composition of both the steels are great different from those of the other AFSs.

## 2. Classification of Hot-Forging Process for Low and Medium Carbon Steels

The hot-forging process of various low and medium carbon forging steels is classified as Table 1. The forging process can be divided into three categories such as conventional hot-forging, controlled hot-forging (or semi-hot-forging), and ausforming. Conventional hot-forging temperatures are between 1000 °C and 1200 °C above the recrystallized temperature ($T_R$). Controlled forging temperatures are between $T_R$ and $Ac_3$. Ausforming temperatures are lower than $Ac_3$ corresponding to a metastable austenite region. The lower the forging temperature, the finer the microstructure. For every forging, post cooling rates decide the type of matrix structure, namely, quenching, rapid cooling, and moderate cooling resulting in martensite, bainite, and ferrite-pearlite matrix structures, respectively. Microalloying of V, Ti, and/or Nb contributes to the refining of prior austenitic grain and the precipitation strengthening of the matrix structure by V,Ti,Nb(C,N)-carbonitrides [8,38–45]. Microalloying of Cr, Mo, Ni, B, etc. increases the hardenability and resultantly enhances the yield stress and tensile strength of the forging products.

**Table 1.** Classification of hot-forging process for low and medium carbon steels. PHFP steel: V-microalloying precipitation hardening ferritic-pearlitic steel, PHFP-M steel: modified PHFP steel, Q&T steel: quenching and tempering steel, TBF steel: TRIP-aided bainitic ferrite steel, Q&P steel: quenching and partitioning steel, Nano-B steel: nanostructured bainitic steel, TBM steel: TRIP-aided bainitic ferrite-martensitic steel, TM steel: TRIP-aided martensitic steel, M-Mn steel: medium Mn steel, $T_R$: recrystallized temperature.

| Designation | Forging Temperature | Phase on Forging | Steel Type after Hot-Forging | | |
| --- | --- | --- | --- | --- | --- |
| | | | Diffusion Transformation | Non-Diffusion and Diffusion Transformation | Non-Diffusion Transformation |
| Conventional hot-forging | $>T_R$ | austenite | PHFP steel, PHFP-M steel, bainitic steel, TBF steel, one-step Q&P steel, Nano-B steel | Two-step Q&P steel, TBM steel | Q&T steel, TM steel, Martensite type M-Mn steel |
| Controlled hot-forging | $T_R - Ac_3$ | austenite | | | |
| Ausforming | $<Ac_3$ | metastable austenite | | | |

## 3. Conventional High-Strength Forging Steels; CFSs

Figure 2 shows the typical hot-forging and subsequent controlling cooling diagrams of the CFSs such as Q&T, PHFP-M, and bainitic steels [46]. In this case, the temperature, reduction strain, and reduction strain rate on hot-forging and post cooling rate are important hot-forging parameters controlling the microstructure and mechanical properties of the steels. The PHFP-M and bainitic steels [44,45] have made great achievements for weight reduction of hot-forging components in the project "Lightweight Forging Initiative" performed in Germany [1,2]. The mechanical properties and typical steel grade of the PHFP-M and bainitic steels are summarized as follows, together with Q&T steels.

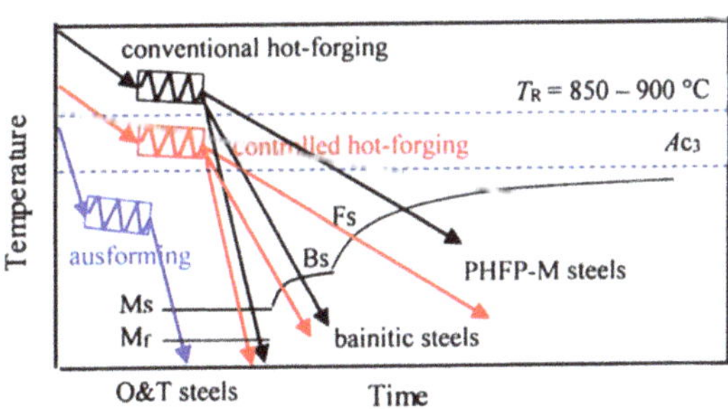


**Figure 2.** Conventional hot-forging, controlled hot-forging, and ausforming process followed by controlling cooling for the CFSs such as Q&T, bainitic, and PHFP-M steels. $T_R$: recrystallized temperature. Tempering is carried out after quenching in Q&T steel. (The figure of Ref. 46 is modified).

### 3.1. Q&T Steels

Conventional Q&T steels used in many hot-forged parts achieve great mechanical properties such as high yield stress, high tensile strength, large uniform and total elongations, and high impact toughness, as well as high fatigue strength, although the Q&T treatment is expensive. When the tensile strength and impact toughness are compared with those of other CFSs, Q&T steel possesses the highest tensile strength and impact toughness (Figure 3) [3,4,47]. Typical Q&T steel grade DIN-25CrMo4 and 42CrMo4 are applied to bolts and screws [4,48,49], which must be strong and tough to undergo the cyclic load. If low-cost is required for the materials, DIN-36CrB4 without Mo is replaced for the DIN-42CrMo4 [48,49], also with regard to applications such as crankshafts and shafts. Fully substituting Mo by Mn results in reduced low temperature impact toughness values [48].

Ausforming of metastable austenite is an important forging technique enhancing both the strength and impact toughness due to prior austenitic grain refining. A martensitic METT100 steel (0.07C–3.2Mn–0.6Cr–0.2Mo, in masss%) [50] has been developed by Q&T treatment after ausforming. The METT100 steel has high buckling strength two times that of DIN-S40VC (0.40C–0.2Si–1.0Mn–(0.1–0.2)V). The increased yield stress, tensile strength, and impact toughness are caused by the increased dislocation density without cell structure and refining of martensite lath and block, as well as prior austenitic grain refining [50]. The METT100 forging steel is expected to be applied to actual automotive components such as connecting rod, crank shaft, driveshaft differential, constant velocity joint, wheel carrier, suspension, etc.

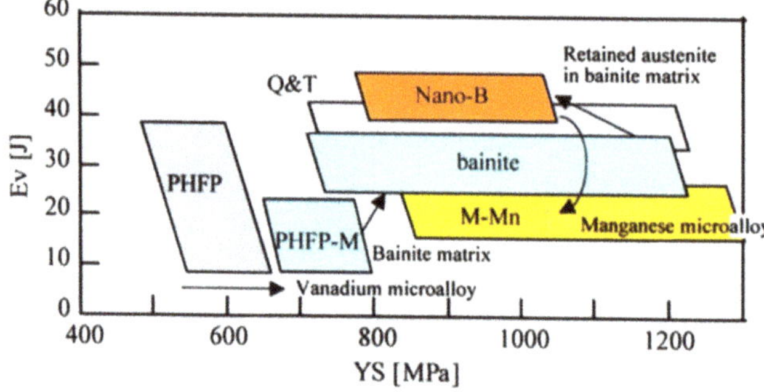


**Figure 3.** Relationship between Charpy V-notch impact energy (*Ev*) and yield stress (YS) at room temperature in various steel groups, mainly used for hot-forgings in the automotive industry [3,4,47]. Q&T, PHFP, PHFP-M, Nano-B, and M-Mn are quenching and tempering steel, V-microalloying precipitation-hardening ferritic-pearlitic steel, modified PHFP steel, nanostructured bainitic steel, and dual-phase type medium Mn steel, respectively.

### 3.2. PHFP-M Steels

PHFP-M steels developed as alternative materials of Q&T forging steels are characterized by low production cost due to the elimination of an additional Q&T step. The PHFP-M steels are achieved by reduction of the ferrite fraction, the decrease in the pearlite lamellae spacing, and the addition of the microalloying elements Nb and Ti, which results in additional precipitates besides the V(C,N) [3,38–40,45,51,52]. However, the PHFP-M steels possess lower yield strength, lower tensile strength, and lower Charpy V- notch impact energy compared to the Q&T steels, although the yield stress and tensile strength are higher than those of the PHFP steels with 0.1 to 0.4 mass% V (Figure 3). The typical steel grade is DIN-38MnVS6 [41,48] and 46MnVS5 [53] for connecting rods. The fine V,Nb,Ti-carbonitrides which were precipitated by fast cooling to about 600 °C followed by holding at the temperature [4] effectively increase the yield stress and tensile strength, with the decreased elongations and impact toughness.

Controlled hot-forging increases the yield stress and tensile strength of the PHFP-M forging steels due to the prior austenitic grain refining (and refining of ferritic and pearlitic structures), although the ferrite fraction increases [46]. In this case, forging temperatures are between 800 °C and 1000 °C. The controlled hot-forging is applied to produce METT75 steel (0.38C−0.25Si−1.0Mn−0.2Cr−0.2V) and METT80 steel (0.35C−0.25Si−1.0Mn−0.2Cr−0.3V) [54]. These steels are produced via air cooling after hot-forging, resulting in very fine VC precipitates in the ferrite phase. The non-heat-treated METT80 steel achieves high yield strength 1.6 times that of conventional S40VC steel [54]. For applications to powertrain components, ultrahigh-strength PHFP-M steel (Vanard Ultra: (0.4−0.5)C−(0.3−0.8)Si−(1−1.5)Mn−(0.15−0.25)V−(0.01−0.025)N) is also developed by Clarke et al. [55].

### 3.3. Bainitic Steels

Bainitic steels have an impact on the toughness−yield stress relationship intermediate between Q&T and PHFP-M steels (Figure 3) [49,56−61]. The bainitic steels with the different bainite morphologies such as acicular, upper, and lower bainites can be formed simply by controlling the post cooling rate immediately after hot-forging (Figure 2). One bainitic steel grade DIN-20MnCrMo7 is commercially available for applications to common rail and injector body [48,53,56,60]. The desired bainitic structure is achieved by adding Mn, Cr, and some Mo. Likewise, DIN-16MnCrV7-7, which is a cost-effective steel grade, achieves an attractive strength level without additional heat-treatment [48]. An important aspect of the bainitic steels is their machinability. For DIN-20MnCrMo7 with 0.15 mass% S, Biermann et al. [58] compared the machinability with that of Q&T steel DIN-42CrMo4. The bainitic steel is more difficult to machine, mainly due to its higher hardness.

Sourmail et al. [59,61] design the medium carbon bainitic forging steel grade DIN-38MnCrMoVB5 and 40MnSiCrMoB4. The benefit of the bainitic steels against Q&T steel DIN-42CrMo4 is obvious since the materials never require heat-treatment after hot-forging [48]. To further improve the mechanical properties of the bainitic steels, the development of Si bearing Nano-B steels with high ductility and high toughness continues [3,4,16,40,53,62] (Figure 3). Simultaneously, many researchers are developing dual-phase type M-Mn steels with a large amount of metastable retained austenite. Unfortunately, the toughness of the M-Mn steels is lower than bainitic steels [47,53,62] (see M-Mn steel in Figure 3). The details of the Nano-B steels are described in the following Section 4.

## 4. Advanced High-Strength Forging Steels; AFSs

Recently, 980−1960 MPa grade AFSs such as TBF, Nano-B, one and two-step Q&P, TBM, TM, and martensite type M-Mn forging steels with bainitic ferrite and/or martensite matrix structure have been developed for weight reduction of automotive powertrain and chassis [32−36]. These prospective AFSs contain Si and/or Al higher than 0.5 mass% to suppress the carbide formation and promote a predominant formation of carbon-enriched retained austenite [32−37]. In addition, IT process at $T_{IT}$ higher than $M_S$, between $M_S$ and $M_f$ and lower than $M_f$ immediately after hot-forging must be conducted to the AFSs (Figure 4). The heat-treatment of two-step Q&P steel exceptionally consists of quenching to temperature ($T_Q$) between $M_S$ and $M_f$ and subsequent partitioning at temperature ($T_P$) higher than $M_S$. In some cases, Cr, Ni, Mo, B, etc. are microalloyed to increase the hardenability of the AFSs. Further, V, Ti, and/or Nb are added to refine the prior austenitic grain. The microstructures of the AFSs are classified into three types as illustrated in Figure 5. Metastable retained austenite in all AFSs plays an important role in enhancing the ductility and fracture strengths such as fatigue strength, impact toughness, and delayed fracture strength [63,64]. In the following, the microstructural and mechanical properties of three types of AFS are detailed.

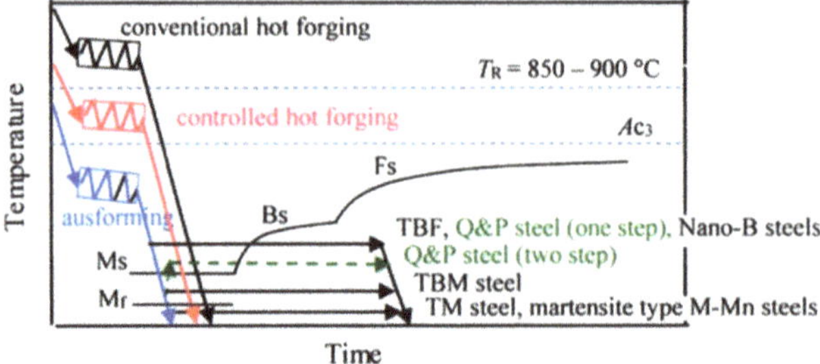

**Figure 4.** Hot-forging processes of low and medium carbon AFSs such as TBF, one-step and two-step Q&P, Nano-B, TBM, TM, and martensite type M-Mn steels. $T_R$: recrystallized temperature.

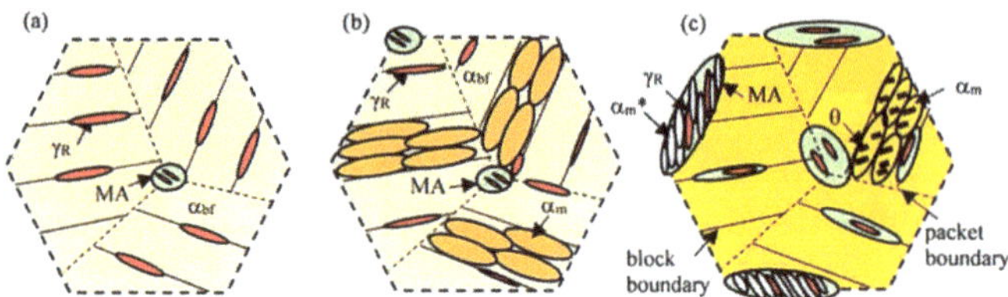

**Figure 5.** Illustration of typical microstructure of various AFSs. (**a**): TBF, Nano-B, and one-step Q&P steels, (**b**): two-step Q&P, TBM, and Nano-B steels, (**c**): TM and martensite type M-Mn steels. $\alpha_{bf}$, $\alpha_m$, $\alpha_m{}^*$, $\gamma_R$, $\theta$, and MA represent bainitic ferrite, soft martensite, hard martensite, retained austenite, carbide, and a mixture of martensite and austenite, respectively.

### 4.1. TBF, One-Step Q&P, and Nano-B Steels ($T_{IT} > M_S$)

TBF steels are produced by IT process at $T_{IT}$ higher than $M_S$ immediately after hot-forging in austenite region, in the same way as one-step Q&P and Nano-B steels (Figure 4). The microstructure mainly consists of bainitic ferrite matrix structure, filmy retained austenite along the bainitic ferrite lath boundary, and a negligible amount of MA phase (Figure 5a).

Hot-forging with a reduction strain higher than 40% refines the prior austenitic grain, bainitic ferrite structure, retained austenite phase, and MA phase (Figure 6b) [37]. Also, the hot-forging increases the volume fraction and mechanical stability of the retained austenite. Microalloying of Cr and Mo is effective to increase the volume fraction and carbon concentration of the retained austenite, as well as an increase in hardenability [32–34]. Controlled hot-forging and ausforming followed by IT process is also further effective to refine the microstructure and improve the retained austenite characteristics [34].

Hot-forging with a reduction strain of 50% achieves an excellent combination of yield stress of 700–1000 MPa and Charpy impact absorbed value of 110–130 J/cm$^2$ in 0.20C–1.52Si–1.50Mn–0.05Nb–0.0018B and 0.42C–1.47Si–1.51Mn–0.50Cr–0.20Mo–0.48Al–0.05Nb TBF steels subjected to IT process at $T_{IT}$ above $M_S$ (Figure 7) [35,65], as well as a good balance of yield stress and total elongation. The yield stress–impact toughness balance exceeds so much that of a hot-forged 0.3C–0.26Si–1.0Mn–0.2Cr–0.3V PHFP-M steel [55] and an ausformed 0.13C–0.26Si–2.7Mn Q&T steel [50] (see Figure 7). A similar result is also obtained by El-Din et al. [66]. From the examinations of microstructure and retained austenite characteristics, it is revealed that the excellent balance of TBF steel is mainly caused by (i) refined bainitic ferrite matrix structure, (ii) refined prior austenitic grain, and (iii) a large amount of refined metastable retained austenite. The retained austenite suppresses the crack initiation or the void formation and subsequent void coalescence during impact tests via the plastic relaxation of localized stress concentration and an increase in carbon-enriched hard martensite fraction resulting from the strain-induced transformation [32,33]. Ultra-fast heating before hot-forging is also effective for grain refining [67].

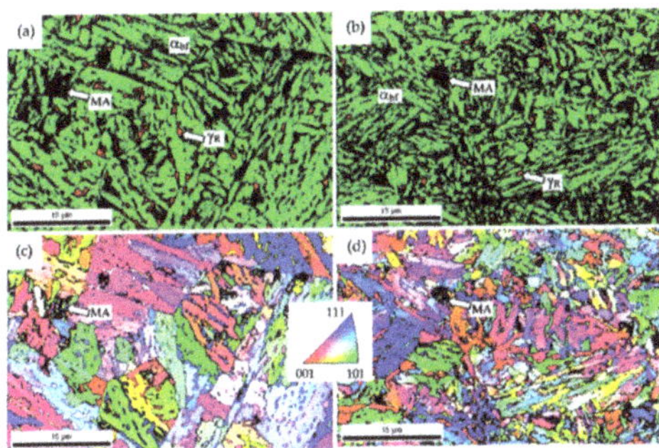

**Figure 6.** Phase maps (**a**,**b**) and orientation maps (**c**,**d**) of EBSP of 0.42C−1.47Si−1.51Mn−0.50Cr−0.20Mo−0.48Al−0.05Nb TBF steel subjected to IT process at $T_{IT}$ = 350 °C immediately after (**a**,**c**) austenitizing and (**b**,**d**) hot-forging at a reduction strain of 50% and a strain rate of 0.5/s. In (**a**,**b**), $\alpha_{bf}$, $\gamma_R$ and MA represent bainitic ferrite (yellowish green), retained austenite (red), and martensite-austenite phase, respectively [32,33]. (**a**,**c**): $f\gamma_0$ = 20.0 vol.%, (**b**,**d**): $f\gamma_0$ = 21.2 vol.%.

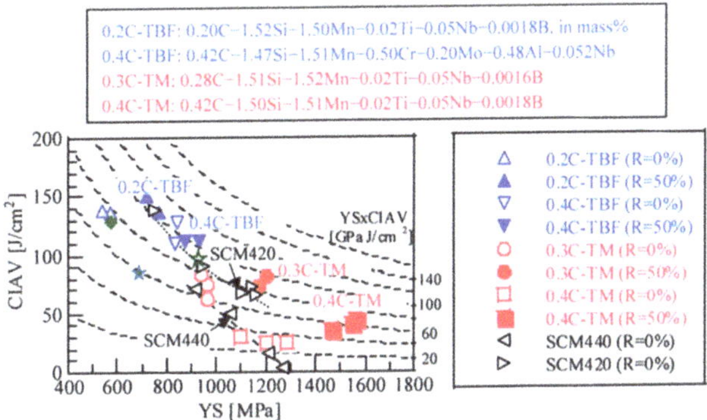


**Figure 7.** Charpy impact absorbed values (CIAV) as a function of yield stress or 0.2% offset proof stress (YS) in hot-forged low and medium carbon TBF and TM steels and heat-treated JIS-SCM420 and JIS-SCM440 steels [33,35,65]. Charpy impact and tensile tests were carried out at room temperature. ☆: 0.2C−1.5Si−5Mn martensite type M-Mn steel [26], ◆: 0.3C−0.26Si−1.0Mn−0.2Cr−0.3V PHFP-M steel [55], ★: 0.13C−0.26Si−2.7Mn ausformed Q&T steel [50]. *R*: reduction strain, open marks: steels without hot-forging, solid marks: steels hot-forged at a reduction strain of 50% and a strain rate of 0.5/s.

Sugimoto et al. [63,68] report that a low carbon TBF steel achieves the excellent balance of mechanical properties as shown in Figure 8. Wirths et al. [69] find out an interesting result which hot-forged 0.18C−0.97Si−2.50Mn−0.2Cr−0.1Mo−0.0018B Nano-B steel exhibits a significant cyclic hardening, differing from 42CrMo4 Q&T steel showing a large cyclic softening, in the same way as a 0.17C−1.41Si−2.02Mn TBF steel [63]. The Nano-B steel also exhibits a balance of tensile strength and Charpy impact absorbed value higher than that of Q&T steel DIN-42CrMo4. Buchmayr [4] and Caballero et al. [23] also report that low carbon Nano-B steels achieve an excellent balance of yield stress and total elongation.

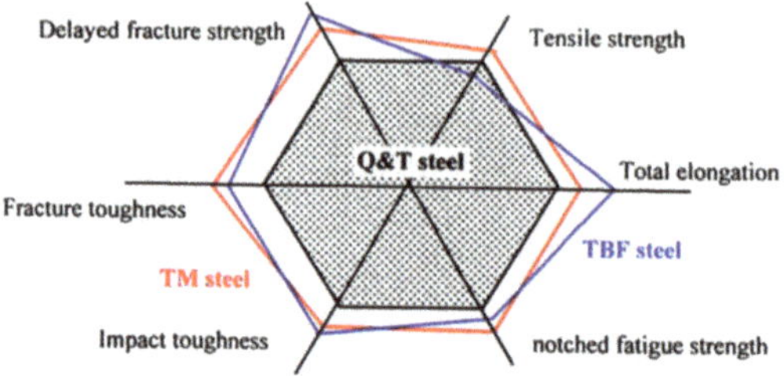



**Figure 8.** Comparison of various mechanical properties of DIN-22MnB5 Q&T steel and 0.2C-1.5Si-1.5Mn TBF and TM steels subjected to heat-treatment without hot-forging. This figure was modified based on Refs. [63,64].

### 4.2. Two-Step Q&P ($M_f < T_Q < M_S$, $T_P > M_S$) and TBM Steels ($M_f < T_{IT} < M_S$)

Two-step Q&P process consists of direct quenching to temperature ($T_Q$) between $M_S$ and $M_f$ and subsequent partitioning at temperature ($T_P$) higher than $M_S$ after austenitizing [17] (green dotted line in Figure 4). On quenching, a certain amount of soft martensite transforms first. The soft martensite fraction ($f\alpha_m$) can be estimated by the following empirical equation proposed by Koistinen and Marburger [70].

$$f\alpha_m = 1 - \exp\left\{-A\,(M_S - T_Q)^B\right\} \tag{1}$$

where A and B are material constants. During partitioning, carbide-free bainite transformation results from austenite, accompanied with carbon migration from soft martensite to the remaining austenite. The resultant microstructure consists of a dual-phase structure of soft martensite and bainitic ferrite and a large amount of metastable retained austenite (Figure 5b). In some cases, a small quantity of MA phase is formed [71]. It is noteworthy that fine martensite in the MA phase is a very hard phase because it is carbon-enriched to the same extent as retained austenite. A small amount of carbide is formed only in the soft martensite lath structure when the quenching temperature is slightly higher than $M_f$ [72]. A simplified route with direct cooling in the quenching bath for making hot-forged parts using a Q&P process is illustrated in Figure 9 [45].

The two-step Q&P process brings out a balance of tensile strength and total elongation higher than one-step Q&P process [17]. Gao et al. [73,74] report that the two-step Q&P process enhances the Charpy impact absorbed values in low carbon Si–Mn steels. According to Bagliani et al. [72] and Somani et al. [75], the two-step Q&P process lowers the ductile–brittle transition temperature of Charpy impact absorbed value in a 0.28C−1.41Si−0.67Mn−1.49Cr−0.56Mo steel and (0.19−0.22)C−(0.55−1.48)Si−(1.50−2.04)Mn−(0−1.06)Al−(0.52−1.20)Cr−(0−0.21)Mo−(0−0.79)Ni steels, respectively. Also, De Diego-Calderón et al. [76] confirm that the two-step Q&P process increases the fracture toughness of a 0.25C−1.5Si−3Mn steel.

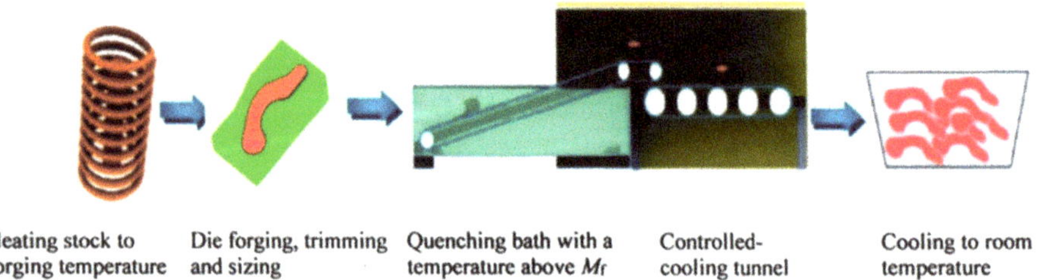



**Figure 9.** Simplified route with direct cooling in quenching bath for making forged parts using a two-step Q&P process (Reprinted with permission from [44], Copyright IRED 2014).

Sugimoto et al. [32,65] also reported that excellent ductility and impact toughness are obtained in hot-forged low and medium carbon TBM steels. They propose that the excellent mechanical properties are associated with (i) a refined dual-phase matrix structure of soft martensite and bainitic ferrite, (ii) refined prior austenitic grain, and (iii) a large amount of refined metastable retained austenite. The dual-phase structure plays a role in decreasing the fracture facet size and consequently lowers the ductile–brittle transition temperature. In addition, the dual-phase structure produces a compressive internal stress in the bainitic ferrite structure. A similar study using a $0.15C–1.41Si–1.88Mn–1.88Cr–0.36Ni–0.34Mo$ Nano-B steel subjected to ausforming and then IT process at $T_{IT}$ between $M_S$ and $M_f$ is reported by Zhao et al. [77]. The roles of the above (ii) and (iii) are the same as those of TBF steels.

Although there are no data comparing the mechanical properties of hot-forged two-step Q&P steels with those of hot-forged TBM steels, the mechanical properties of the hot-forged two-step Q&P steels are supposed to be the same extent as those of the hot-forged TBM steels [64] because their microstructures are nearly the same as each other.

*4.3. TM and Martensite Type M-Mn Steels ($T_{IT} < M_f$)*

TM steel can be produced by IT process at $T_{IT}$ lower than $M_f$ immediately after hot-forging of austenite, as well as direct quenching to room temperature (Figure 4) [12–16]. If the $T_{IT}$ is near room temperature or direct quenching to room temperature is conducted, partitioning at $T_P$ lower than $M_S$ is added. The microstructure is characterized by a dual-phase structure of soft martensite matrix structure and hard MA phase (Figure 5c). Most of a small amount of retained austenite is included in the MA phases [12–16]. Only a little carbide is developed only in the soft martensite, in the same way as two-step Q&P quenched to $T_Q$ just higher than $M_f$ and TBM steel subjected to IT process at $T_{IT}$ just higher than $M_f$. Final partitioning promotes carbon-enrichment into the retained austenite and softening of both martensites without additive carbide formation [35–37].

Hot-forging at a reduction of 50% brings on an excellent combination of tensile strength of 1500 to 2000 MPa (or yield stress of 1200 to 1560 MPa) and Charpy impact absorbed value of 35 to 80 J/cm$^2$ in 0.3C- and 0.4C-TM steels when the partitioning process was added after the IT process (Figure 7) [35]. The combination exceeds so much those of the commercial Q&T steels such as JIS-SCM420 and SCM440 steels in a range of YS > 1200 MPa [35,65], although the combination is slightly inferior to that of TBF steels. The ductile–brittle transition temperatures, however, are much lower than those of TBF steels [64]. Also, TM steels possess higher tensile strength, delayed fracture strength, fracture toughness, and notched fatigue strength than TBF steel. As shown in Figure 7a, 0.2C–1.5Si–5Mn martensite type M-Mn steel without hot-forging exhibits the same combination of yield stress and Charpy impact absorbed value as 0.3C-TM steel, but lower combination than 0.2C- and 0.4C-TBF steels [26].

According to Kobayashi et al. [78], MA phases in the TM steel play an important role in suppressing void formation and preferential void growth at the MA phase/matrix interface (Figure 10a). Furthermore, the MA phases also inhibit the initiation and propagation of cleavage cracking (Figure 10b), through the block effect and the plastic relaxation that occurs as a result of the strain-induced transformation of the metastable retained austenite. With this is in mind, the excellent impact toughness of TM steel is mainly caused by (i) refined dual-phase structure of soft martensite and MA phase and (ii) refined prior austenitic grain [35,36]. The (i) also generates a high long range compressive internal stress in the soft martensitic matrix. A small quantity of metastable retained austenite and the decreased carbide fraction may make a contribution to the impact toughness.

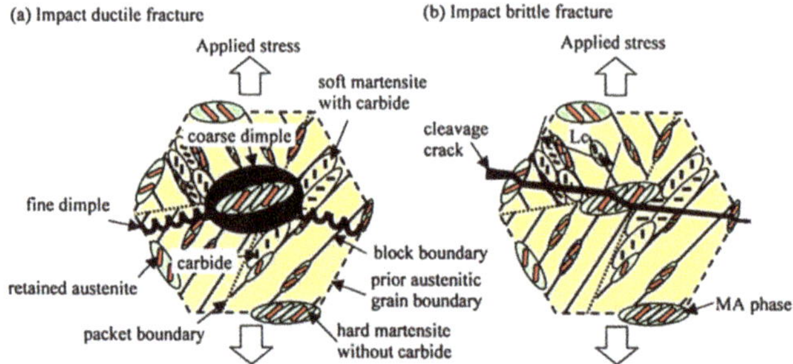



**Figure 10.** Illustrations showing (**a**) a ductile fracture and (**b**) a brittle fracture of TM steel appeared on impact tests [78,79]. $L_c$: Quasi-cleavage length affected by the MA phase located on prior austenitic, packet, and block boundaries.

## 5. Summary

In this paper, the hot-forging process and mechanical properties of CFSs were first stated, as well as typical steel grade and various automotive applications. Next, the microstructural and mechanical properties of HFSs were shown and compared with those of CFSs.

Hot-forged AFSs achieved much better mechanical properties than hot-forged CFSs. The excellent mechanical properties, especially impact toughness, were mainly caused by the following microstructural properties in TBF steel (and one-step Q&P and Nano-B steels), two-step Q&P steel (and TBM steel), and TM steel (and martensite type M-Mn steel).

(1)   TBF, one-step Q&P, and Nano-B steels: refined bainitic ferrite matrix structure, refined prior austenitic grain, and a large amount of refined metastable retained austenite.
(2)   Two-step Q&P and TBM steels: refined dual-phase structure of soft martensite and bainitic ferrite, refined prior austenitic grain, and a large amount of refined metastable retained austenite.
(3)   TM and martensite type M-Mn steels: refined dual-phase structure of soft martensite and MA phase and refined prior austenitic grain.

For impact toughness, two types of dual-phase structure play roles in enhancing a compressive internal stress in soft structure and decreasing the fracture facet size. A large amount of refined metastable retained austenite suppresses the crack initiation or the void formation and subsequent void coalescence or crack growth via the plastic relaxation of localized stress concentration resulting from the strain-induced transformation of the retained austenite. The strain-induced hard martensite also plays a role in the block effect of crack growth, as well as MA phase. Although TM steel includes a small quantity of retained austenite, the metastable retained austenite makes a contribution to high impact toughness, as well as the decreased carbide.

The AFSs can be expected to enable the weight reduction and size-down of the automotive powertrain and chassis parts. In order to apply the AFSs to the automotive hot-forging components, however, other mechanical properties such as fatigue strength, delayed fracture strength, and wear properties of hot-forged AFSs must be systematically investigated in the future. In addition, it is desired that their mechanical properties are compared to those of CFSs and the other AFSs such as dual-phase type M-Mn steels [52,80,81] and TWIP steels [7–10].

To further enhance the wear resistance and fatigue strength of the AFSs, a surface layer shell hardened up to 60 HRC is needed [48,79]. Many engineers also want to know such mechanical properties of case-hardened AFSs.

Finally, authors want to emphasize that the AFSs may be applied to not only automotive powertrain and chassis parts but also the forging parts of other engineering structures such as construction machinery, airplanes, marine machinery, etc. The applications will increase many fracture strengths and resultantly bring about a great increase in reliability.

**Author Contributions:** The first draft of the paper has been written by K.-i.S., T.H. and A.K.S., and the final editing has been done by K.-i.S. and A.K.S.

**Funding:** This research received no external funding.

**Acknowledgments:** We thank J. Kobayashi from Ibaraki University for their kind discussion.

**Conflicts of Interest:** The authors declare no conflict of interest.

## References


1. Raedt, H.W.; Wilke, F.; Ernst, C.S. Lightweight forging initiative—Phase II: Lightweight design potential in a light commercial vehicle. *ATZ* **2016**, *118*, 49–52. [CrossRef]
2. Raedt, H.W.; Wurm, T.; Busse, A. The lightweight forging initiative—Phase III: Lightweight forging design for hybrid cars and heavy-duty trucks. *ATZ* **2019**, *120*, 54–59. [CrossRef]
3. Keul, C.; Wirths, V.; Bleck, W. New bainitic steel for forgings. *Arch. Civ. Mech. Eng.* **2012**, *12*, 119–125. [CrossRef]
4. Buchmayr, B. Critical assessment 22: Bainitic forging steels. *Mater. Sci. Technol.* **2016**, *32*, 517–522. [CrossRef]
5. Zackay, V.F.; Parker, E.R.; Fahr, D.; Bush, R. The enhancement of ductility in high-strength steels. *Trans. Am. Soc. Met.* **1967**, *60*, 252–259.
6. De Cooman, B.C. Structure-properties relationship in TRIP steels containing carbide-free bainite. *Curr. Opinion Solid State Mater. Sci.* **2004**, *8*, 285–303. [CrossRef]
7. Bleck, W.; Guo, X.; Ma, Y. The TRIP effect and its application in cold formable sheet steels. *Steel Res. Int.* **2017**, *88*, 1700218. [CrossRef]
8. Rana, R.; Singh, S.B. *Automotive Steels—Design, Metallurgy, Processing and Applications*; Woodhead Publishing: Cambridge, UK, 2016.
9. Brux, U.; Frommeyer, G.; Grassel, O.; Meyer, L.W.; Weise, A. Development and characterization of high strength impact resistant Fe-Mn-(Al-,Si) TRIP/TWIP steels. *Steel Res. Int.* **2002**, *73*, 294–298. [CrossRef]
10. De Cooman, B.C.; Estrin, Y.; Kim, S.K. Twinning-induced plasticity (TWIP) steels. *Acta Mater.* **2018**, *142*, 283–362. [CrossRef]
11. Sugimoto, K.; Murata, M.; Song, S. Formability of Al-Nb bearing ultrahigh-strength TRIP-aided sheet steels with bainitic ferrite and/or martensite matrix. *ISIJ Int.* **2010**, *50*, 162–168. [CrossRef]
12. Sugimoto, K.; Kobayashi, J. Newly developed TRIP-aided martensitic steels. In Proceedings of the Materials Science and Technology and Exhibition 2010 (MS&T 10), Houston, TX, USA, 17–21 October 2010; pp. 1639–1649.
13. Kobayashi, J.; Pham, D.V.; Sugimoto, K. Stretch-flangeability of 1.5 GPa grade TRIP-aided martensitic cold rolled sheet steels. *Steel Res. Int. (Spec. Ed.)* **2011**, *81*, 598–603.
14. Kobayashi, J.; Song, S.; Sugimoto, K. Ultrahigh-strength TRIP-aided martensitic steels. *ISIJ Int.* **2012**, *52*, 1124–1129. [CrossRef]
15. Sugimoto, K.; Kobayashi, J.; Pham, D.V. Advanced ultrahigh-strength TRIP-aided martensitic sheet steels for automotive applications. In Proceedings of the International Symposium on New Developments in Advanced High Strength Sheet Steels (AIST 2013), Vail, CO, USA, 23–27 June 2013; pp. 175–184.
16. Pham, D.V.; Kobayashi, J.; Sugimoto, K. Effects of microalloying on stretch-flangeability of TRIP-aided martensitic sheet steel. *ISIJ Int.* **2014**, *54*, 1943–1951. [CrossRef]
17. Speer, J.G.; Edmonds, D.V.; Rizzo, F.C.; Matlock, D.K. Partitioning of carbon from supersaturated plates of ferrite with application to steel processing and fundamentals of the bainitic transformation. *Curr. Opin. Solid State Mat. Sci.* **2004**, *8*, 219–237. [CrossRef]
18. Kähkönen, J.; Pierce, D.T.; Speer, J.G.; De Moor, E.; Thomas, G.A.; Coughlin, D.; Clarke, K.; Clarke, A. Quenched and partitioned CMnSi steels containing 0.3 wt.% and 0.4 wt.% carbon. *JOM* **2016**, *68*, 210–214. [CrossRef]
19. Wang, L.; Speer, J.G. Quenching and partitioning steel heat treatment. *Metallogr. Microstruc. Anal.* **2013**, *2*, 268–281. [CrossRef]
20. Bhadeshia, H.K.D.H. Some phase transformations in steels. *Mater. Sci. Technol.* **1999**, *15*, 22–29. [CrossRef]
21. Garcia-Mateo, C.; Caballero, F.G.; Sourmail, T.; Smanio, V.; De Andres, C.G. Industrialised nanocrystalline bainitic steels—Design approach. *Int. J. Mater. Res.* **2014**, *105*, 725–734. [CrossRef]

22. Zhang, X.; Xu, G.; Wang, X.; Embury, D.; Bouaziz, O.; Purdy, G.P.; Zurob, H.S. Mechanical behavior of carbide-free medium carbon bainitic steels. *Metall. Mater. Trans. A* **2014**, *45A*, 1352–1361. [CrossRef]
23. Caballero, F.G.; Santofimia, M.J.; Garcia-Mateo, C.; Chao, J.; Garcia de Andres, C. Theoretical design and advanced microstructure in super high strength steels. *Mater. Des.* **2009**, *30*, 2077–2083. [CrossRef]
24. Cao, W.; Wang, C.; Shi, J.; Wang, M.; Hui, W.; Dong, D. Microstructure and mechanical properties of Fe–0.2C–5Mn steel processed by ART-annealing. *Mater. Sci. Eng. A* **2011**, *528*, 6661–6666. [CrossRef]
25. Seo, E.; Cho, L.; De Cooman, B.C. Application of quenching and partitioning processing to medium Mn Steel. *Metall. Mater. Trans. A* **2015**, *46A*, 27–31. [CrossRef]
26. Sugimoto, K.; Tanino, H.; Kobayashi, J. Impact toughness of medium-Mn transformation-induced plasticity-aided steels. *Steel Res. Int.* **2015**, *86*, 1151–1160. [CrossRef]
27. Raabe, D.; Ponge, D.; Dmitriva, O.; Sander, B. Nanoprecipitate-hardened 1.5 GPa steels with unexpected high ductility. *Scr. Mater.* **2009**, *60*, 1141–1144. [CrossRef]
28. Image IT Media. Nissan. Available online: http://image.itmedia.co.jp/l/im/mn/articles/1303/14/l_sp_130314nissan_03.jpg (accessed on 1 October 2019).
29. Billur, E.; Cetin, B.; Gurleyik, M. New generation advanced high strength steels: Developments, Trends and Constraints. *Int. J. Sci. Technol. Res.* **2016**, *2*, 50–62.
30. Yoshioka, N.; Tachibana, M. Weight reduction of automotive seat components using high-strength steel with high formability. *Kobe Steel Eng. Rep.* **2017**, *66*, 22–25.
31. Wang, W.; Wang, H.; Liu, S. Development of hot-rolled Q&P steel at Baosteel. In Proceedings of the 9th Pacific Rim International Conference on Advanced Materials and Processing (PRICM9), Kyoto, Japan, 1–5 August 2016.
32. Sugimoto, K.; Sato, S.; Arai, G. Hot forging of ultrahigh-strength TRIP-aided steel. *Mater. Sci. Forum* **2010**, *638–642*, 3074–3079. [CrossRef]
33. Sugimoto, K.; Sato, S.; Kobayashi, J.; Srivastava, A.K. Effects of Cr and Mo on mechanical properties of hot-forged medium carbon TRIP-aided bainitic ferrite steels. *Metals* **2019**, *9*, 1066. [CrossRef]
34. Sugimoto, K.; Itoh, M.; Hojo, T.; Hashimoto, S.; Ikeda, S.; Arai, G. Microstructure and mechanical properties of ausformed ultrahigh-strength TRIP-aided steels. *Mater. Sci. Forum* **2007**, *539–543*, 4309–4314. [CrossRef]
35. Kobayashi, J.; Sugimoto, K.; Arai, G. Effects of hot forging process on combination of strength and toughness in ultrahigh-strength TRIP-aided martensitic steels. *Adv. Mater. Res.* **2012**, *409*, 696–701. [CrossRef]
36. Kobayashi, J.; Nakajima, Y.; Sugimoto, K. Effects of cooling rate on impact toughness of an ultrahigh-strength TRIP-aided martensitic steel. *Adv. Mater. Res.* **2014**, *922*, 366–371. [CrossRef]
37. Sugimoto, K.; Kobayashi, J.; Nakajima, Y.; Kochi, T. The effects of cooling rate on retained austenite characteristics of a 0.2C-1.5Si-1.5Mn-1.0Cr-0.05Nb TRIP-aided martensitic steel. *Mater. Sci. Forum* **2014**, *783–786*, 1015–1020. [CrossRef]
38. Matlock, D.K.; Speer, J.G. Microalloying concepts and application in long products. *Mater. Sci. Technol.* **2009**, *25*, 1118–1125. [CrossRef]
39. Hui, W.; Xiao, N.; Zhao, X.; Zhang, Y.; Wu, Y. Effect of vanadium on dynamic continuous cooling transformation behavior of medium-carbon forging steels. *J. Iron Steel Res. Int.* **2017**, *24*, 641–648. [CrossRef]
40. Garcia-Mateo, C.; Morales-Rivas, L.; Caballero, F.G.; Milbourn, D.; Sourmail, T. Vanadium effect on a medium carbon forging steel. *Metals* **2016**, *6*, 130. [CrossRef]
41. Li, Y.; Milbourn, D. Vanadium microalloyed forging steel. In Proceedings of the 2nd International Symposium on Automobile Steel, Anshan, China, 21–24 May 2013; pp. 47–54.
42. Miyamoto, G.; Hori, R.; Poorganji, B.; Furuhara, T. Interface precipitation of VC and resultant hardening in V-added medium carbon steels. *ISIJ Int.* **2011**, *51*, 1733–1739. [CrossRef]
43. Hui, W.; Zhang, Y.; Shao, C.; Chen, S.; Zhao, X.; Dong, H. Enhancing the mechanical properties of vanadium-microalloyed medium-carbon steel by optimizing post-forging cooling conditions. *Mater. Manuf. Process.* **2016**, *31*, 770–775. [CrossRef]
44. Mašek, B.; Jirková, H. New material concept for forging. In Proceedings of the International Conference on Advances in Civil Structural Mechanical Engineering (ACSME 2014), Bangkok, Thailand, 4–5 January 2014; pp. 92–96.
45. Balart, M.J.; Davis, C.L.; Strangwood, M. Fracture behavior in medium-carbon Ti-V-Nb and V-Nb microalloyed ferritic-pearlitic and bainitic forging steels with enhanced machinability. *Mater. Sci. Eng. A* **2002**, *328*, 48–57. [CrossRef]

46. Fujiwara, M.; Yoshida, H.; Isogawa, S. Controlled forging technology for special steels. *Denki-Seiko* **2007**, *78*, 259–265. [CrossRef]

47. Gramlich, A.; Emmrich, R.; Bleck, W. Austenite reversion tempering-annealing of 4 wt. % manganese steels for automotive forging application. In Proceedings of the 4th International Conference on Medium and High Mn Steels, Aachen, Germany, 1–3 April 2019; pp. 283–286.

48. Raedt, H.W.; Speckenheuer, U.; Vollrath, K. New forged steels, Energy-efficient solutions for stronger parts. *ATZ* **2012**, *114*, 5–9.

49. Riedner, S.; Van Soest, F.; Kunow, S. B-alloyed quench and temper steels—An option to classic CrMo and CrNiMo quench and temper steels. In Proceedings of the 3rd International Conference on Steels for Cars and Trucks (SCT 2011), Salzburg, Austria, 5–9 June 2011; pp. 219–225.

50. Yoshida, H.; Isogawa, S.; Ishikawa, T. Basic property of martensitic microalloyed steel—Development of microalloyed steel for forging using ausforming I. *J. Jpn. Soc. Technol. Plast.* **2000**, *41*, 379–383.

51. Bleck, W.; Bambach, M.; Wirths, V.; Stieben, A. Microalloyed engineering steels with improved performance—An Overview. *Heat Treat. Mater. J.* **2017**, *72*, 355–364. [CrossRef]

52. Langeborg, R.; Sandberg, O.; Roberts, W. *Fundamentals of Microalloying Forging Steels*; Krauss, G., Banerji, S.K., Eds.; TMS: Warrendale, PA, USA, 1987; pp. 39–54.

53. Dickert, H.H.; Florian, J.C.; Gervelmeyer, J. Air-hardening high strength steels for all dimensions (of forged components). In Proceedings of the 5th International Conference on Steels for Cars and Trucks (SCT2017), Amsterdam, The Netherland, 18–22 June 2017; p. 212.

54. Yoshida, H.; Isogawa, S.; Ishikawa, T. Basic properties of ferrite-pearlitic microalloyed steel—Development of microalloyed steel for controlled forging I. *J. Jpn. Soc. Technol. Plast.* **2001**, *41*, 569–573.

55. Clarke, P.; Green, M.; Dolman, R. Development of high strength microalloyed steel for powertrain applications. In Proceedings of the 4th International Conference on Steels for Cars and Trucks (SCT2014), Braunschweig, Germany, 15–19 June 2014; pp. 442–449.

56. Engineer, S.; Justinger, H.; Janßen, O.; Härtel, M.; Hampel, C.; Randelhoff, T. Technological properties of the new high strength bainitic steel 20MnCrMo7. In Proceedings of the 3rd International Conference on Steels for Cars and Trucks (SCT 2011), Salzburg, Austria, 5–9 June 2011; pp. 404–411.

57. Gomez, G.; Pérez, T.; Bhadeshia, H.K.D.H. Air cooled bainitic steels for strong, seamless pipes Part 1—Alloy design, kinetics and microstructure. *Mater. Sci. Technol.* **2009**, *25*, 1502–1507. [CrossRef]

58. Biermann, D.; Felderhoff, F.; Engineer, S.; Justinger, H. Machinability of high-strength bainitic steel 20MnCrMo7. In Proceedings of the 3rd International Conference on Steels for Cars and Trucks (SCT 2011), Salzburg, Austria, 5–9 June 2011; pp. 471–478.

59. Sourmail, T.; Smanio, V.; Galtier, A.; Marchal, F. Bainitic steel grades for forged components: Role of transformation kinetics and design for induction hardened components. In Proceedings of the 4th International Conference on Steels for Cars and Trucks (SCT2014), Braunschweig, Germany, 15–19 June 2014; pp. 342–349.

60. Roelofs, H.; Krull, H.-G.; Lembke, M.; Lapierre Boire, L.P. Continuously cooled high strength steels: From small to large sections. In Proceedings of the 5th International Conference on Steels for Cars and Trucks (SCT2017), Amsterdam, The Netherland, 18–22 June 2017; p. 82.

61. Maminska, K.; Roth, A.; d'Eramo, E.; Marchal, F.; Galtier, A.; Sourmail, T. A new bainitic forging steel for surface induction hardened components. In Proceedings of the 5th International Conference on Steels for Cars and Trucks (SCT2017), Amsterdam, The Netherland, 18–22 June 2017; p. 228.

62. Bleck, W.; Prahl, U.; Hirt, G.; Bambach, M. Designing new forging steels by ICMPE. In *Advances in Production Technology*; Brecher, C., Ed.; Springer: Cham, Switzerland; Heidelberg, Germany, 2015; Chapter 7; pp. 85–98.

63. Sugimoto, K. Fracture strength and toughness of ultrahigh strength TRIP aided steels. *Mater. Sci. Technol.* **2009**, *25*, 1108–1117. [CrossRef]

64. Sugimoto, K.; Srivastava, A.K. Microstructure and Mechanical Properties of a TRIP-aided Martensitic Steel. *Metall. Microstr. Anal.* **2015**, *4*, 344–354. [CrossRef]

65. Sugimoto, K.; Kobayashi, J.; Arai, G. Development of low alloy TRIP-aided steel for hot-forging parts with excellent toughness. *Steel Res. Int. (Spec. Ed. Metal Form. 2010)* **2010**, *81*, 254–257.

66. El-Din, H.N.; Showaib, E.A.; Zaafarani, N.; Refaiy, H. Structure-properties relationship in TRIP type bainitic ferrite steel austempered at different temperatures. *Int. J. Mech. Mater. Eng.* **2017**, *12*, 3. [CrossRef]

67. Papaefthymiou, S.; Banis, A.; Bouzouni, M.; Petrov, R.H. Effect of ultra-fast heat treatment on the subsequent formation of mixed martensitic/bainitic microstructure with carbides in a CrMo medium carbon steel. *Metals* **2019**, *9*, 312. [CrossRef]
68. Yoshikawa, N.; Kobayashi, J.; Sugimoto, K. Notch-fatigue properties of advanced TRIP-aided bainitic ferrite steels. *Metall. Mater. Trans. A* **2012**, *43A*, 4129–4136. [CrossRef]
69. Wirths, V.; Bleck, W. Carbide free bainitic forging steels with improved fatigue properties. In Proceedings of the 4th International Conference on Steels for Cars and Trucks (SCT2014), Braunschweig, Germany, 15–19 June 2014; pp. 350–357.
70. Koistinen, D.P.; Marburger, R.E. A general equation prescribing the extent of the austenite-martensite transformation in pure iron-carbon alloys and plain carbon steels. *Acta Metal.* **1959**, *7*, 59–60. [CrossRef]
71. Kobayashi, J.; Ina, D.; Yoshikawa, N.; Sugimoto, K. Effects of the addition of Cr, Mo and Ni on the microstructure and retained austenite characteristics of 0.2%C-Si-Mn-Nb ultrahigh-strength TRIP-aided bainitic ferrite steels. *ISIJ Int.* **2012**, *52*, 1894–1901. [CrossRef]
72. Bagliani, E.P.; Santofimia, M.J.; Zhao, L.; Sietsma, J.; Anelli, E. Microstructure, tensile and toughness properties after quenching and partitioning treatments of a medium-carbon steel. *Mater. Sci. Eng. A* **2013**, *559*, 486–495. [CrossRef]
73. Gao, G.; An, B.; Zhang, H.; Guo, H.; Gui, X.; Bai, B. Concurrent enhancement of ductility and toughness in an ultrahigh strength lean alloy steel treated by bainite-based quenching-partitioning-tempering process. *Mater. Sci. Eng. A* **2017**, *702*, 104–122. [CrossRef]
74. Gao, G.; Zhang, H.; Gui, X.; Luo, P.; Tan, Z.; Bai, B. Enhanced ductility and toughness in an ultrahigh-strength Mn-Si-Cr-C steel: The great potential of ultrafine filmy retained austenite. *Acta Mater.* **2014**, *76*, 425–433. [CrossRef]
75. Somani, M.C.; Porter, D.A.; Karjalainen, L.P.; Suikkanen, P.P.; Misra, R.D.K. Process design for tough ductile martensitic steels through direct quenching and partitioning. *Mater. Today Proc.* **2015**, *2* (Suppl. 3), S631–S634. [CrossRef]
76. De Diego-Calderón, I.; Sabirov, I.; Molina-Aldareguia, J.M.; Föjer, C.; Thiessen, R. Microstructural design in quenched and partitioned (Q&P) steels to improve their fracture properties. *Mater. Sci. Eng. A* **2016**, *657*, 136–146.
77. Zhao, L.; Qian, L.; Zhou, Q.; Li, D.; Wang, T.; Jia, Z.; Zhang, F.; Meng, J. The combining effects of ausforming on the transformation kinetics, microstructure and mechanical properties of low-carbon bainitic steel. *Mater. Des.* **2019**, *183*, 108123. [CrossRef]
78. Kobayashi, J.; Ina, D.; Nakajima, Y.; Sugimoto, K. Effects of microalloying on the impact toughness of ultrahigh-strength TRIP-aided martensitic steel. *Metall. Mat. Trans. A* **2013**, *44A*, 5006–5017. [CrossRef]
79. Sugimoto, K.; Hojo, T.; Srivastava, A.K. An overview of fatigue strength of case-hardening TRIP-aided martensitic steels. *Metals* **2018**, *8*, 355. [CrossRef]
80. Tanino, H.; Horita, M.; Sugimoto, K. Impact toughness of 0.2 Pct C-1.5 Pct Si-(1.5 to 5) Pct Mn transformation-induced plasticity-aided steels with an annealed martensite matrix. *Metall. Mater. Trans. A* **2016**, *47A*, 2073–2080. [CrossRef]
81. Gramlich, A.; Emmrich, R.; Bleck, W. Austenite reversion tempering-annealing of 4 wt.% manganese steels for automotive forging application. *Metals* **2019**, *9*, 575. [CrossRef]

*Article*

# Effects of Thermomechanical Processing on Hydrogen Embrittlement Properties of UltraHigh-Strength TRIP-Aided Bainitic Ferrite Steels

Tomohiko Hojo [1,*], Yutao Zhou [1,2], Junya Kobayashi [3], Koh-ichi Sugimoto [4], Yoshito Takemoto [5], Akihiko Nagasaka [6], Motomichi Koyama [1], Saya Ajito [1] and Eiji Akiyama [1]

[1] Institute for Materials Research, Tohoku University, Sendai 980-8577, Japan; zhou.yutao.s3@dc.tohoku.ac.jp (Y.Z.); motomichi.koyama.c5@tohoku.ac.jp (M.K.); saya.ajito.d1@tohoku.ac.jp (S.A.); eiji.akiyama.d1@tohoku.ac.jp (E.A.)

[2] Graduate School of Engineering, Tohoku University, Sendai 980-8579, Japan

[3] College of Engineering, Ibaraki University, Hitachi 316-8511, Japan; junya.kobayashi.jkoba@vc.ibaraki.ac.jp

[4] School of Science and Technology, Shinshu University, Nagano 380-8553, Japan; sugimot@shinshu-u.ac.jp

[5] Graduate School of National Science and Technology, Okayama University, Okayama 700-8530, Japan; tanutake@okayama-u.ac.jp

[6] National Institute of Technology, Nagano College, Nagano 381-8550, Japan; nagasaka@nagano-nct.ac.jp

* Correspondence: tomohiko.hojo.a1@tohoku.ac.jp; Tel.: +81-22-215-2062

**Abstract:** The effects of thermomechanical processing on the microstructure and hydrogen embrittlement properties of ultrahigh-strength, low-alloy, transformation-induced plasticity (TRIP)-aided bainitic ferrite (TBF) steels were investigated to apply to automobile forging parts such as engine and drivetrain parts. The hydrogen embrittlement properties were evaluated by conducting conventional tensile tests after hydrogen charging and constant load four-point bending tests with hydrogen charging. The 0.4 mass%C-TBF steel achieved refinement of the microstructure, improved retained austenite characteristics, and strengthening, owing to thermomechanical processing. This might be attributed to dynamic and static recrystallizations during thermomechanical processing in TBF steels. Moreover, the hydrogen embrittlement resistances were improved by the thermomechanical processing in TBF steels. This might be caused by the refinement of the microstructure, an increase in the stability of the retained austenite, and low hydrogen absorption of the thermomechanically processed TBF steels.

**Keywords:** TRIP-aided bainitic ferrite steel; thermomechanical processing; hydrogen embrittlement; retained austenite

**Citation:** Hojo, T.; Zhou, Y.; Kobayashi, J.; Sugimoto, K.-i.; Takemoto, Y.; Nagasaka, A.; Koyama, M.; Ajito, S.; Akiyama, E. Effects of Thermomechanical Processing on Hydrogen Embrittlement Properties of UltraHigh-Strength TRIP-Aided Bainitic Ferrite Steels. *Metals* **2022**, *12*, 269. https://doi.org/10.3390/met12020269

Academic Editor: Marcello Cabibbo

Received: 22 December 2021
Accepted: 28 January 2022
Published: 31 January 2022

**Publisher's Note:** MDPI stays neutral with regard to jurisdictional claims in published maps and institutional affiliations.

## 1. Introduction

Ultrahigh-strength steel sheets with a tensile strength of 980 MPa and greater have been applied to automobile structural parts to reduce the weights of vehicles and improve collision safety [1]. Thus, in recent years, the mechanical properties and fracture morphologies of high-strength steels were positively investigated [2,3]. In ultrahigh-strength steel sheets, low-alloy transformation-induced plasticity (TRIP)-aided steels [4,5] with a bainitic ferrite matrix (TBF steels) [6] are expected to be the next-generation advanced high-strength steels (AHSS), owing to their high strength associated with a relatively high dislocation density in the matrix and excellent fatigue [7,8] and impact [9,10] properties due to the TRIP effect of retained austenite. The conventional TBF steels were produced by annealing at the austenite region and austempering treatment using cold-rolled steel sheets with chemical compositions of low and medium carbon and an adequate amount of Si and Mn to obtain a microstructure consisting of a bainitic ferrite matrix and retained austenite [6]. On the other hand, downsizing and weight reduction of automobile forging parts such as the engine and drivetrain parts are also required. To resolve these requirements, the TBF steels are expected to be applied in automobile forging parts, because TBF steels possess the abovementioned excellent properties.

The microstructure evolution of high-strength steels during hot and cold rolling and cooling from an austenitizing temperature were investigated [11,12]. Zhao et al. [13] reported that the refinement and strengthening of advanced high-strength steels were achieved by thermomechanical processing. Sugimoto et al. [14] reported that the refined, recrystallized ferrite in an annealed martensite matrix was obtained when hot forging was conducted on the TRIP-aided annealed martensitic steel at an inter-critical annealing temperature. Moreover, Sugimoto et al. [15–17] reported that strengthening, improved retained austenite characteristics, and improved impact properties were achieved via hot and warm forging of TBF steels. The authors listed in [18,19] reported the effects of hot and warm forging on the microstructure evolution, retained austenite characteristics, and mechanical properties of TRIP-aided martensitic (TM) steels and discussed those mechanisms.

On the other hand, hydrogen embrittlement [20,21], which reduces the ductility of high-strength steels with a tensile strength of more than 980 MPa, also becomes a serious problem, similar to conventional high-strength structural steels. Sojka et al. [22] and Laureys et al. [23] reported that conventional TRIP-aided steels with a tensile strength of 780 MPa were sensitive to hydrogen embrittlement, and fracture morphologies were characterized. The authors [24,25] revealed the following three facts in hydrogen-charged TBF steels. First, the hydrogen embrittlement susceptibility increased with a decreasing strain rate. Second, a plastic strain of 3–10% decreased the hydrogen embrittlement susceptibility. Third, hydrogen-related cracks were initiated at the transformed martensite or bainitic ferrite matrix and transformed martensite interfaces. Furthermore, the effects of the alloying elements on the hydrogen embrittlement of TM steels were investigated, and the addition of alloying elements such as Cr was found to improve the hydrogen embrittlement resistance [26]. However, the hydrogen embrittlement behavior of thermomechanically processed TBF steels has not yet been fully elucidated.

In this study, the hydrogen embrittlement resistance of hot-forged TBF steels was evaluated using a tensile testing technique at a conventional strain rate and a constant load four-point bending technique. The former shows that deformation induced transformation of the retained austenite, unlike the latter. In addition, the effects of thermomechanical processing on the microstructural evolution and hydrogen embrittlement properties of TBF steels were investigated.

## 2. Materials and Methods

Hot-rolled steel bars with a diameter of 32 mm and with chemical compositions as listed in Table 1 were prepared in this study. Hereafter, the steels with carbon contents of 0.2, 0.3, and 0.4 mass% are named steels A, B, and C, respectively. The hot-forging specimens with dimensions of 20 mm in height and 90 mm in length were machined from these steel bars. Hot-forged steels A, B and C were produced via annealing at 930 °C for 1200 s, followed by one-step forging at a 50% reduction ratio ($R$) using a 500-ton press machine (Hydraulic press, Amino Corporation, Fujinomiya, Japan) and subsequent austempering at 350 °C for 1000 s using hot-forging specimens, as depicted in Figure 1. Conventional TBF steels that did not undergo hot forging were produced with $R$ = 0% via annealing at 930 °C for 1200 s, followed by austempering at 350 °C for 1000 s to compare the effect of thermomechanical processing. Tensile, four-point bending, microstructure observation, and X-ray diffraction specimens were cut from the hot-forged and heat-treated specimens parallel to the rolling direction at a quarter region in the thickness direction of the hot-forged samples.

The microstructure was observed and analyzed using a scanning electron microscope (SEM Merlin, Zeiss, Oberkochen, Germany) equipped with an electron backscatter diffraction (EBSD) system (OIM Data Collection, OIM-Analysis, TSL solutions, Sagamihara, Japan) operated at an accelerated voltage of 20 kV. The EBSD analyses were conducted in an area of 60 μm × 30 μm with a step size of 0.1 μm. The samples for microstructure observation were ground by waterproof papers of #320 and #600 and were polished using polycrystalline diamond slurries of 9 and 3 μm for the particle size and colloidal silica, respectively.

**Table 1.** Chemical compositions (mass%) of steels.

| Steels | C | Si | Mn | P | S | Al | Nb | Ti | B | O | N |
|---|---|---|---|---|---|---|---|---|---|---|---|
| A | 0.20 | 1.52 | 1.50 | 0.004 | 0.0021 | 0.039 | 0.05 | 0.02 | 0.0018 | 0.001 | 0.0011 |
| B | 0.28 | 1.51 | 1.52 | <0.005 | 0.0011 | 0.041 | 0.051 | 0.02 | 0.0016 | 0.001 | 0.0012 |
| C | 0.42 | 1.50 | 1.51 | <0.005 | 0.0009 | 0.043 | 0.05 | 0.02 | 0.0018 | 0.0019 | 0.0035 |

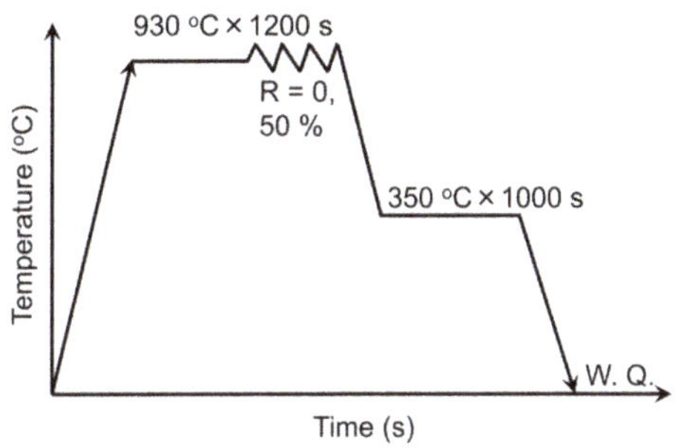


**Figure 1.** Thermomechanical processing and heat treatment diagrams of steels A, B and C. *R* represents reduction ratio. W. Q. indicates water quenching.

The retained austenite characteristics of hot-forged steels A, B, and C were analyzed by the X-ray diffraction (XRD) method using an X-ray diffractometer (Rigaku Co. Ltd., RINT2000, Tokyo, Japan). The volume fraction of the retained austenite ($f_\gamma$) was quantified by the integral intensities of $\alpha$-Fe(200), $\alpha$-Fe(211), $\gamma$-Fe(200), $\gamma$-Fe(220), and $\gamma$-Fe(311) diffraction peaks obtained using CuK$\alpha$ radiation. The carbon concentration in the retained austenite ($C_\gamma$) was estimated using Equation (1) [27] from the average lattice parameter ($a_\gamma$ ($\times 10^{-10}$ m)), which was measured from the $\gamma$-Fe(200), $\gamma$-Fe(220), and $\gamma$-Fe(311) diffraction peaks of the CuK$\alpha$ radiation:

$$a_\gamma = 3.5780 + 0.0330C_\gamma + 0.00095Mn_\gamma + 0.0056Al_\gamma + 0.0220N_\gamma + 0.0051Nb_\gamma + 0.0031Mo_\gamma + 0.0039Ti_\gamma, \tag{1}$$

where $Mn_\gamma$, $Al_\gamma$, $N_\gamma$, $Nb_\gamma$, $Mo_\gamma$, and $Ti_\gamma$ represent the concentrations of the respective individual elements (mass%) in the retained austenite. In this study, the contents of the added alloying elements were substituted for these concentrations.

Tensile tests were carried out with a tensile testing machine (AG X plus 100 kN, Shimadzu Co. Ltd., Kyoto, Japan) at a crosshead speed of 1 mm/min (i.e., an initial strain rate of $8.33 \times 10^{-4}$ /s) at 25 °C with and without hydrogen using tensile specimens with dimensions of 15 mm in gauge length, 6 mm in wide, and 1.2 mm in thickness. The hydrogen embrittlement properties were evaluated by hydrogen embrittlement susceptibility (*HES*), which was calculated using Equation (2) [24,28].

$$HES = (1 - \varepsilon_1 / \varepsilon_0) \times 100\%, \tag{2}$$

where $\varepsilon_0$ and $\varepsilon_1$ denote the total elongation without and with hydrogen, respectively. Hydrogen charging to the tensile specimens was conducted via cathodic charging using a 3 mass% NaCl (Sodium Chloride, FUJIFILM Wako Pure Chemical Co. Ltd., Osaka, Japan) aqueous solution containing 5 g/L NH$_4$SCN (Ammonium Thiocyanate, FUJIFILM Wako Pure Chemical Co. Ltd., Osaka, Japan) at a current density of 10 A/m$^2$ at 25 °C for 48 h before conducting the tensile tests.

Constant load tests were conducted by means of 4-point bending tests using a rectangular specimen with dimensions of 65 mm in length, 10 mm in wide, and 1.2 in mm thickness with hydrogen charging in a 0.5 mol/L $H_2SO_4$ (Sulfuric Acid, FUJIFILM Wako Pure Chemical Co. Ltd., Osaka, Japan) + 0.01 mol/L $NH_4SCN$ (Ammonium Thiocyanate, FUJIFILM Wako Pure Chemical Co. Ltd., Osaka, Japan) solution at 25 °C and a current density of 500 A/$m^2$. The hydrogen embrittlement properties evaluated by the four-point bending tests were defined by the delayed fracture strength (*DFS*), which was the maximum bending stress without failure for 5 h in the specimen.

The hydrogen concentrations in steels A, B, and C were measured by hydrogen thermal desorption analysis (TDA) using gas chromatography. The samples which were charged with hydrogen under the same charging condition as that of the tensile tests were heated between the ambient temperature and 300 °C at a heating rate of 200 °C/h. The diffusible hydrogen concentration was defined as the total hydrogen concentration desorbed between the ambient temperature and 150 °C. After hydrogen charging, the samples were immediately kept in liquid nitrogen to prevent hydrogen desorption. The samples picked up from the cryogenic temperature were rinsed by ultrasonic cleaning using acetone before TDA. The intervals between picking up the sample and the start of TDA were approximately 15 min.

## 3. Results

### 3.1. Microstructure and Tensile Properties

Figure 2 depicts the inverse pole figure (IPF) and phase maps analyzed by EBSD in steels A, B, and C with and without hot forging. In conventional steels A, B, and C without thermomechanical processing, the microstructure of steel A with 0.2 mass%C consisted of a coarse bainitic ferrite matrix and retained austenite, which was located at the packet, block, and lath boundaries. In steels B and C, which exhibited higher carbon contents, the microstructure was characterized as a fine and uniform lath bainitic ferrite matrix and film-type retained austenite located between the packet, block, and lath boundaries. On the other hand, when steels A, B, and C were subjected to thermomechanical processing, the microstructure of steel A was changed to fine ferrite grains with a grain diameter of approximately 10 μm, fine granular retained austenite, and a small amount of bainitic ferrite. In steel B with a carbon content of 0.3 mass%, the fine granular ferrite and bainitic ferrite lath coexisted as a matrix, and fine granular retained austenite was observed at the prior austenitic grain, block, and packet boundaries. In addition, prior austenitic grain refinement and shortened bainitic ferrite laths were achieved, and consequently, a fine and uniform microstructure was achieved by thermomechanical processing in steel C with a carbon content of 0.4 mass%. The film-type retained austenite was located between the shortened bainitic ferrite lath, whereas fine granular retained austenite existed at the fine prior austenitic grain boundaries.

Table 2 lists the retained austenite characteristics and tensile properties of the conventional and hot-forged steels A, B, and C. The initial volume fraction ($f_{\gamma 0}$) and initial carbon concentration ($C_{\gamma 0}$) in the retained austenite of conventional steels A, B, and C were 9.1–17.6 vol% and 0.74–1.22 mass%, respectively, and these increased with the increasing carbon content. The $f_{\gamma 0}$ slightly decreased, and $C_{\gamma 0}$ increased in steel A due to the thermomechanical processing. The $f_{\gamma 0}$ increased and $C_{\gamma 0}$ decreased in steels B and C because of the thermomechanical processing.

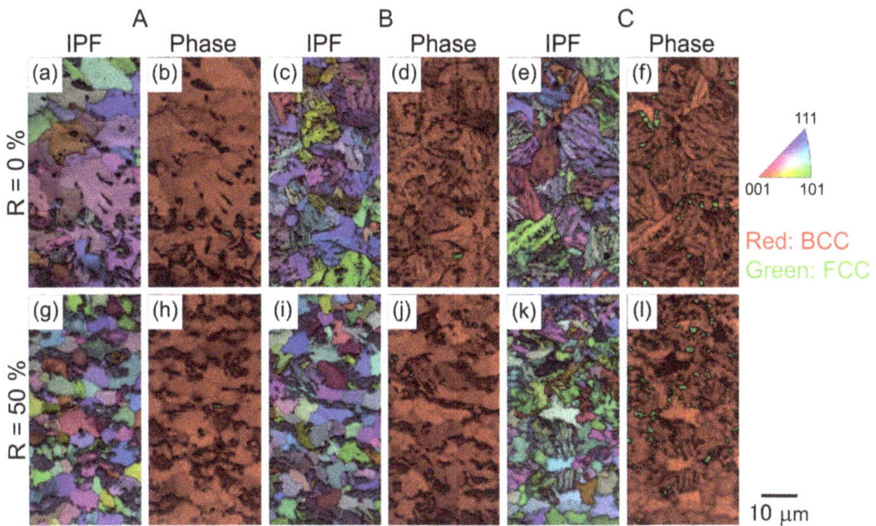

**Figure 2.** (**a,c,e,g,i,k**) Inverse pole figure (IPF) and (**b,d,f,h,j,l**) phase maps of steels (**a,b,g,h**) A, (**c,d,i,j**) B, and (**e,f,k,l**) C (**a–f**) without and (**g–l**) with hot forging. *R* represents reduction ratio. BCC and FCC denote body-centered-cubic and face-centered-cubic, respectively.

**Table 2.** Tensile properties, retained austenite characteristics, and prior austenitic grain sizes of steels A, B and C.

| Steels | *R* | *TS* | *YS* | *YR* | *TEl* | *UEl* | *RA* | *TS* × *TEl* | *HES* | $f_{\gamma 0}$ | $C_{\gamma 0}$ | *d* |
|---|---|---|---|---|---|---|---|---|---|---|---|---|
| A | 0 | 748 | 528 | 0.71 | 32.6 | 21.4 | 39.9 | 24.4 | 9.8 | 9.1 | 0.74 | 10.7 |
|  | 50 | 837 | 500 | 0.60 | 25.4 | 21.5 | 36.2 | 21.3 | 26.4 | 8.9 | 0.98 | 7.4 |
| B | 0 | 950 | 749 | 0.79 | 33.2 | 22.5 | 38.7 | 31.5 | 48.2 | 14.1 | 1.09 | 12.8 |
|  | 50 | 959 | 559 | 0.58 | 26.8 | 21.8 | 32.0 | 25.7 | 34.3 | 14.3 | 0.82 | 7.0 |
| C | 0 | 1097 | 937 | 0.86 | 31.9 | 24.4 | 44.7 | 35.0 | 85.3 | 17.6 | 1.22 | 16.3 |
|  | 50 | 1122 | 788 | 0.71 | 27.6 | 24.1 | 32.2 | 31.0 | 68.5 | 21.1 | 1.01 | 7.3 |

*R* (%): reduction ratio, *TS* (MPa): tensile strength, *YS* (MPa): yield strength, *YR*: yield ratio, *TEl* (%): total elongation, *UEl* (%): uniform elongation, *RA* (%): reduction in area, *TS*×*TEl* (GPa%): strength-ductility balance, *HES* (%): hydrogen embrittlement susceptibility, $f_{\gamma 0}$ (vol%): initial volume fraction of retained austenite, $C_{\gamma 0}$ (mass%): initial carbon concentration in retained austenite, *d* (μm): prior austenitic grain diameter.

Figure 3 depicts the nominal stress–strain curves of the conventional and hot-forged steels A, B, and C without and with hydrogen charging. The tensile properties are shown in Table 2 and Figure 4. The tensile strengths of steels A, B, and C without hydrogen increased from 748 to 837 MPa, from 950 to 959 MPa, and from 1097 to 1122 MPa, respectively, whereas the yield strength decreased from 528 to 500 MPa, from 749 to 559 MPa, and from 937 to 788 MPa, respectively, owing to thermomechanical processing. In addition, thermomechanical processing decreased the total elongation from 32.6 to 25.4%, from 33.2 to 26.8%, and from 31.9 to 27.6%, which corresponded to the reduction ratio in the *TEl* of 22.1%, 19.3%, and 13.5% in steels A, B, and C, respectively, although the uniform elongations of steels A, B, and C without hydrogen were hardly changed via hot forging, and those reduction ratios were −0.5%, 3.1%, and 1.2%, respectively.

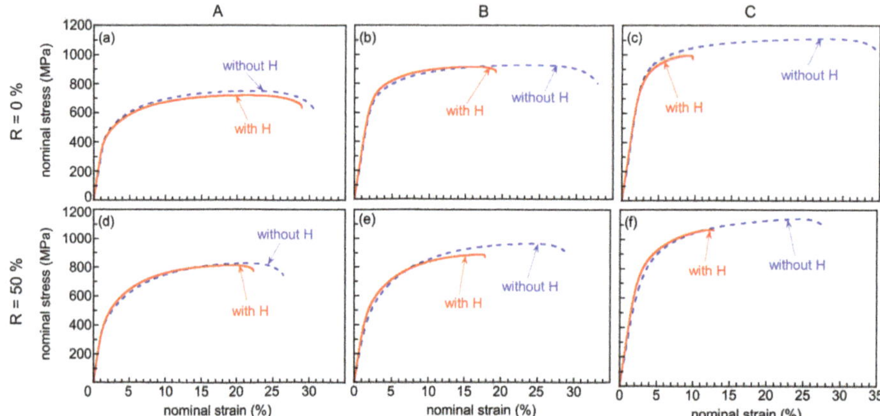

**Figure 3.** Nominal stress–strain curves of (**a,b,c**) conventional and (**d,e,f**) hot-forged steels (**a,d**) A, (**b,e**) B, and (**c,f**) C with and without hydrogen. *R* represents reduction ratio.

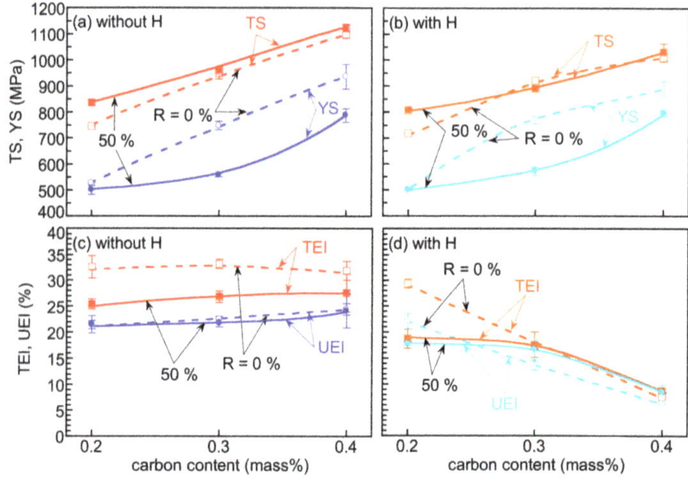

**Figure 4.** Variations in (**a,b**) tensile strength (*TS*), yield strength (*YS*), (**c,d**) total elongation (*TEl*), and uniform elongation (*UEl*) as a function of carbon content in conventional and hot-forged steels A, B, and C (**a,c**) without and (**b,d**) with hydrogen. *R* represents reduction ratio.

### 3.2. Hydrogen Embrittlement Properties Evaluated by Tensile Tests

As depicted in Figure 3, steels A, B, and C exhibited a reduction in the total elongation (fracture elongation) due to hydrogen absorption, although the stress–strain response before the fracture hardly changed. The decrease in the total elongation increased with the increasing carbon content in steels A, B, and C. It should be noted that the tendencies of hydrogen-induced mechanical degradation were similar in both the conventional and thermomechanically processed steels A, B, and C. The typical fracture surfaces of the conventional and hot-forged steels A, B, and C with and without hydrogen are depicted in Figure 5. In the steels without thermomechanical processing without hydrogen, dimples with diameters of approximately 1–5 μm appeared on the fracture surfaces of steels A and B, whereas dimples with area fractions of approximately 5.5% and a quasi-cleavage fracture with that of approximately 94.5% coexisted on the typical fracture surface of steel C. When the conventional steels were charged with hydrogen, the fracture surface of steel

A possessed a mixture of dimples which exhibited dimple diameters of approximately 2–10 μm with area fractions of approximately 23.8% and a quasi-cleavage fracture with that of approximately 76.2%, and those of steels B and C were changed to quasi-cleavage fractures. On the other hand, when the steels were subjected to thermomechanical processing, the fracture surface of steel A with hydrogen charging changed from dimples with diameters of approximately 1.5–6 μm to a mixture of dimples with diameters of approximately 2–4 μm and a quasi-cleavage fracture which possessed area fractions of the dimples of 5.3% and quasi-cleavage of 94.7% in the typical fracture surface. On the other hand, steels B and C exhibited fracture surfaces of a mixture of dimples and quasi-cleavage, regardless of the presence of hydrogen.

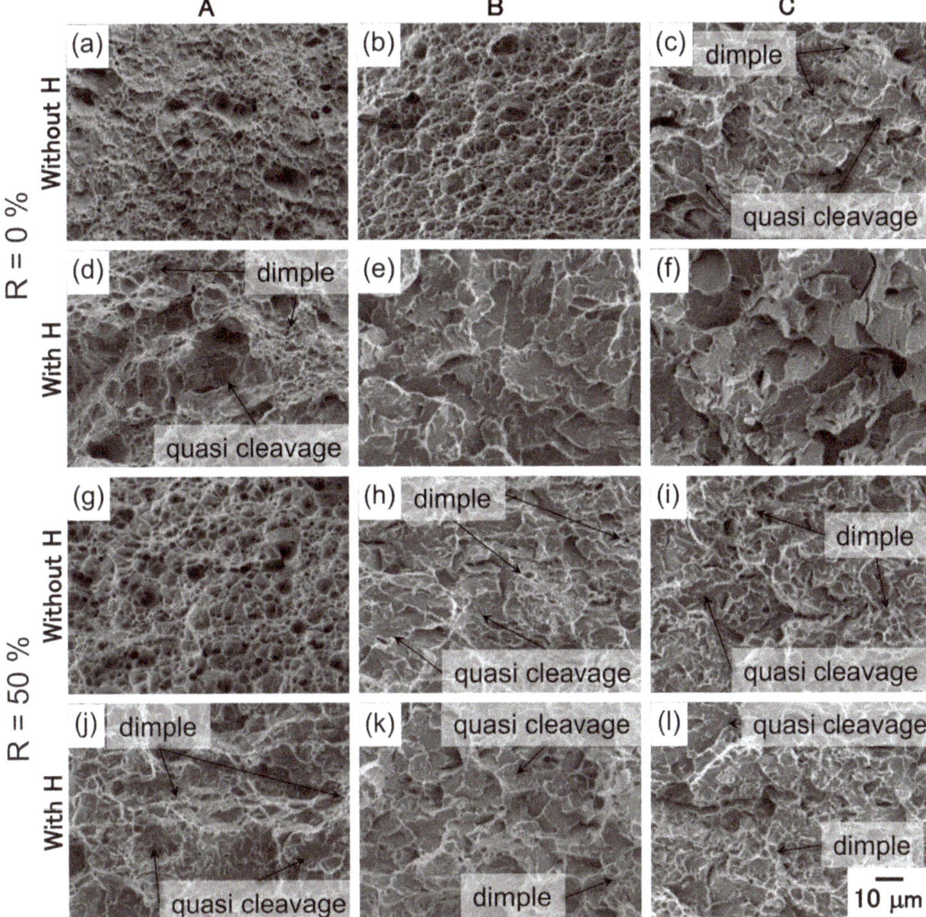


**Figure 5.** Fracture surfaces of (**a–f**) conventional and (**g–l**) hot-forged steels (**a,d,g,j**) A, (**b,e,h,k**) B, and (**c,f,i,l**) C (**d-f,j-l**)with and (**a–c,g-i**) without hydrogen after tensile tests. *R* represents reduction ratio.

Figure 6 depicts the relationships between hydrogen embrittlement susceptibility (*HES*) and tensile strength (*TS*) and yield strength (*YS*) in steels A, B and C. The *HES* increased with increasing *TS* in steels A, B, and C. The thermomechanical processing increased the *HES* of steel A, whereas the *HES* decreased when steels B and C were subjected to hot forging.

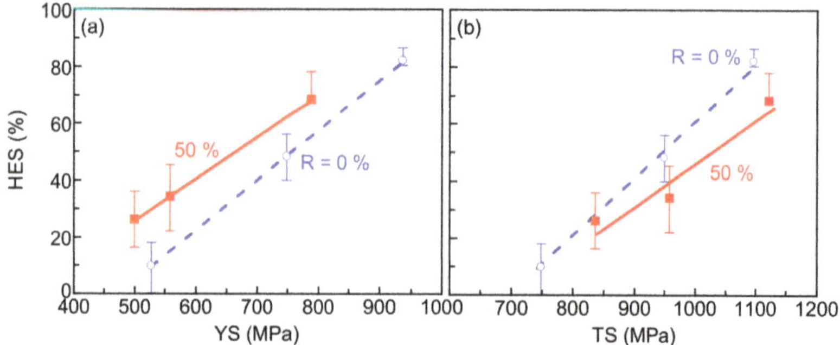


**Figure 6.** Variations in hydrogen embrittlement susceptibility (*HES*) as functions of (**a**) yield strength (*YS*) and (**b**) tensile strength (*TS*) in conventional and hot-forged steels A, B, and C. *R* represents reduction ratio.

Figure 7 depicts the hydrogen desorption curves of the conventional and hot-forged steels A, B, and C. The diffusible hydrogen concentrations obtained from the hydrogen desorption curves in steels A, B, and C are listed in Table 3. It was confirmed that the hydrogen in steels A, B, and C was desorbed between the ambient temperature and approximately 150 °C, and the height of the hydrogen desorption peak increased and its temperature slightly shifted to a higher temperature with an increasing carbon content. The height of the hydrogen desorption peak slightly decreased, and the corresponding diffusible hydrogen concentrations of steels B and C were deduced, although the hydrogen desorption peak temperature did not vary, owing to the thermomechanical processing.

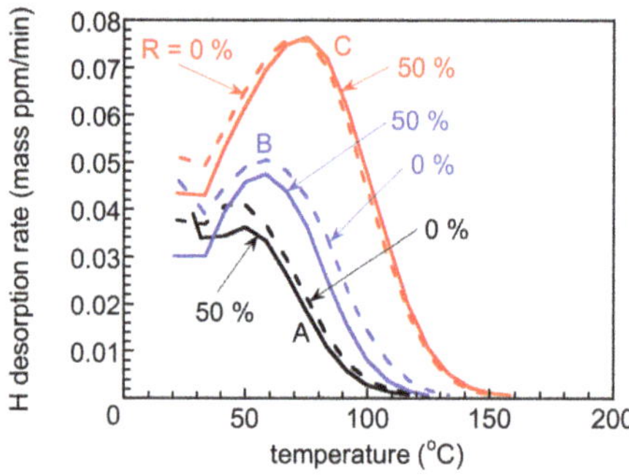


**Figure 7.** Hydrogen desorption curves of conventional and hot-forged steels A, B, and C charged with hydrogen by a 3% NaCl + 5 g/L NH$_4$SCN solution at a current density of 10 A/m$^2$ for 48 h. *R* represents reduction ratio.

**Table 3.** Hydrogen concentration ($H_C$) of conventional and hot-forged steels A, B, and C. *R* represents reduction ratio.

| Steel | R (%) | $H_C$ (Mass ppm) |
|---|---|---|
| A | 0 | 1.36 |
|   | 50 | 1.21 |
| B | 0 | 2.03 |
|   | 50 | 1.64 |
| C | 0 | 3.41 |
|   | 50 | 3.38 |

*3.3. Hydrogen Embrittlement Properties Evaluated by Four-Point Bending Tests*

Figure 8 depicts the applied stress–time to fracture ($\sigma_A$-$t_f$) curves evaluated by the four-point bending tests of the conventional and hot-forged steels A, B, and C. Hydrogen embrittlement did not occur both in steel A with or without thermomechanical processing and in thermomechanically-processed steel B after undergoing four-point bending tests for 5 h. Meanwhile, hydrogen embrittlement occurred in the other steels during the four-point bending tests, and $t_f$ increased with decreasing $\sigma_A$. Figure 9 depicts the relationship between *DFS* and *TS* in the conventional and thermomechanically-processed steels A, B, and C. The *DFS* of steels B and C, which possessed high carbon contents when compared with steel A, improved, owing to the thermomechanical processing. The fracture surfaces of steels B and C after undergoing four-point bending tests are depicted in Figure 10. A quasi-cleavage fracture was observed in the vicinity of the surface of the specimen, where hydrogen embrittlement cracks were initiated in all fractured steels. The fracture surface near the center of the specimen in the thickness direction showed a quasi-cleavage fracture containing flat regions in which facet sizes of 14.1 µm for steel B without thermomechanical processing, 16.4 µm for steel C without thermomechanical processing, and 11.1 µm for steel C with hot forging were similar to the prior austeniti grain or packet sizes of 12.8 µm for steel B without thermomechanical processing, 16.3 µm for steel C without thermomechanical processing, and 7.3 µm for steel C with hot forging, respectively.

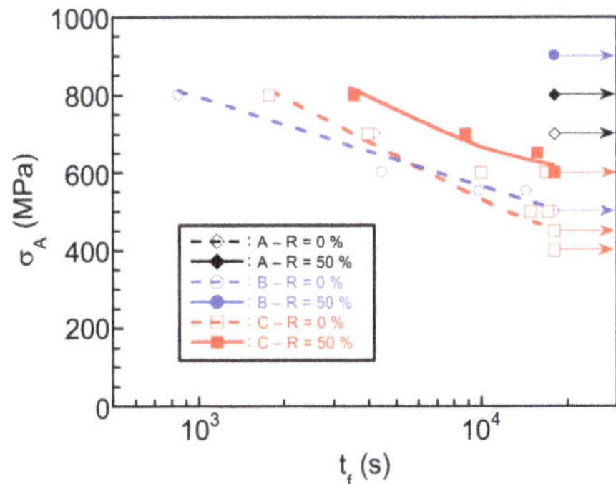



**Figure 8.** Applied stress–time to fracture ($\sigma_A$–$t_f$) curves of conventional and hot-forged steels A, B, and C. *R* represents reduction ratio.

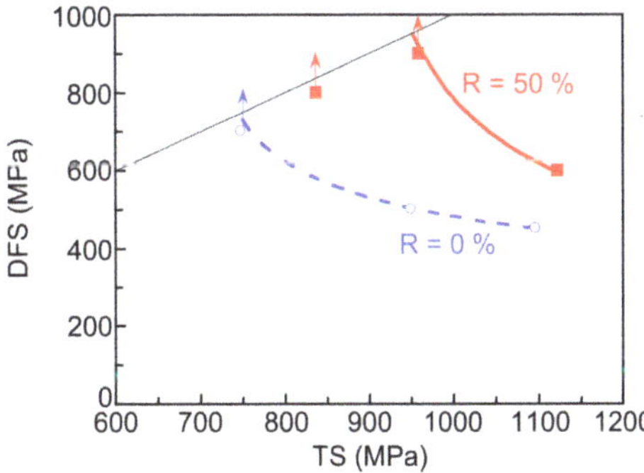



**Figure 9.** Relationship between delayed fracture strength (*DFS*) and tensile strength (*TS*) of conventional and hot-forged steels A, B, and C. *R* represents reduction ratio.

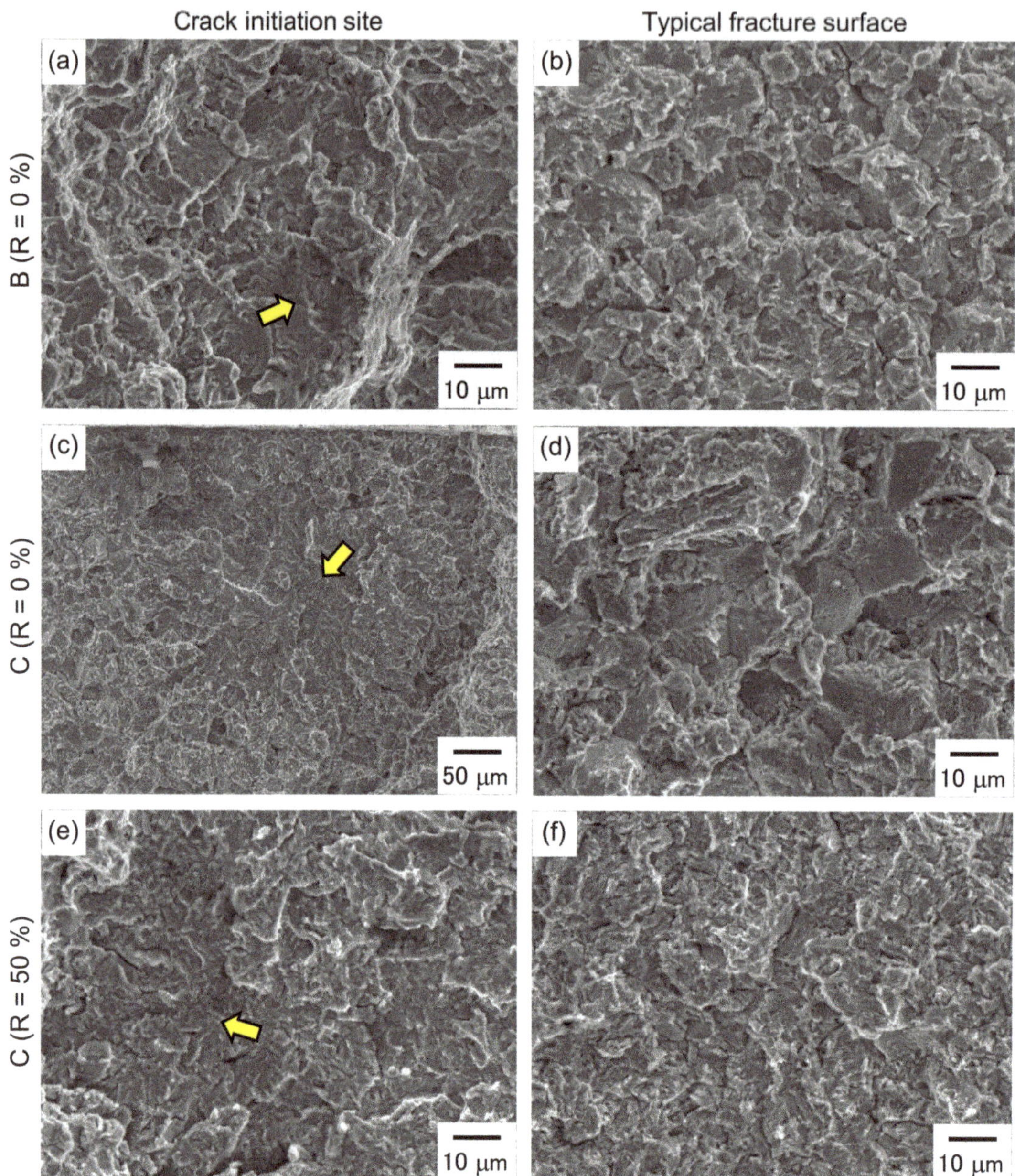

**Figure 10.** Fracture surfaces of (**a**,**b**) conventional steel B and (**c**,**d**) conventional and (**e**,**f**) hot-forged steel C after four-point bending tests, in which arrows represent crack initiation region. *R* represents reduction ratio.

## 4. Discussion

### 4.1. Effects of Thermomechanical Processing on Microstructure Evolution and Mechanical Properties

The microstructure of steel A consisted of fine ferrite grains and a small amount of bainitic ferrite as a matrix and fine granular retained austenite as a second phase, owing to the thermomechanical processing. The microstructural change during the thermomechanical processing of steel A is illustrated in Figure 11. In steel A without hot forging, austenite grains with a diameter of approximately 15–20 µm were obtained during annealing, and prior austenitic grains remained after the bainite transformation (Figure 11b). When steel A was subjected to thermomechanical processing, fine austenite grains were formed, owing to the dynamic or static recrystallizations at the deformation band in the deformed grain and the piled-up dislocations. A large amount of austenite transformed to ferrite during rapid cooling between the austenite region and austempering temperatures (Figure 11c), and bainite transformation of untransformed austenite might occur during austempering because ferrite and bainite transformations are accelerated when compared with conventional heat treatment without hot forging [29,30] (Figure 11e). As a result, steel A exhibited a fine ferrite matrix, retained austenite, and a small amount of bainitic ferrite (Figure 11d). On the other hand, Figure 12 depicts an illustration of the microstructure evolution mechanism of steel C with thermomechanical processing. In steel C, because the dynamic and static recrystallizations occurred during hot forging, fine austenite grains were newly formed in the deformed austenite in the same way as that in the case of steel A (Figure 12c). However, the transformation to ferrite might have rarely occurred because of the high carbon content of 0.4 mass%, although the ferrite transformation was accelerated. The bainite transformation of a large amount of fine austenite occurred during austempering treatment, and the microstructure of hot-forged steel C was composed of a bainitic ferrite matrix with fine block, packet, and lath structures and a large amount of film-type and granular-type retained austenite (Figure 12d).

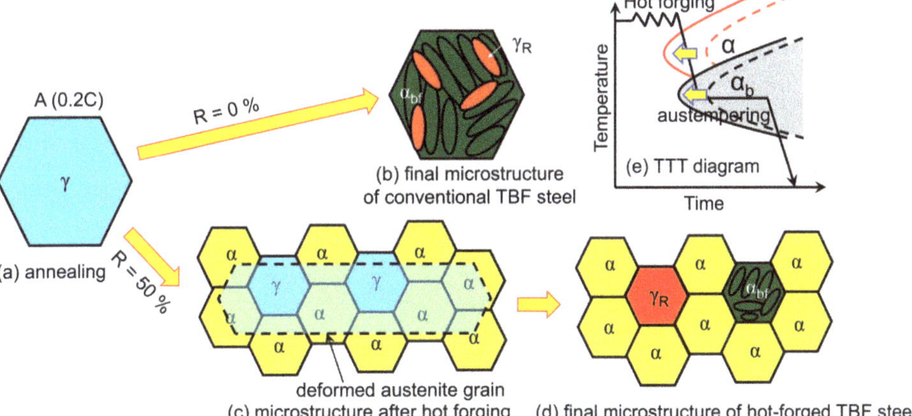


**Figure 11.** Illustrations of microstructure evolution behavior of hot-forged steel A. (**a**) Austenite at annealing temperature, (**b**) final microstructure of conventional TBF steel, (**c**) microstructure after hot forging, (**d**) final microstructure of hot-forged TBF steel, (**e**) TTT diagram, respectively. *R* represents reduction ratio. $\alpha$, $\alpha_b$, $\alpha_{bf}$, $\gamma$, and $\gamma_R$ denote ferrite, bainite, bainitic ferrite, austenite, and retained austenite, respectively. The TTT diagram (**e**) is the time transition temperature diagram.

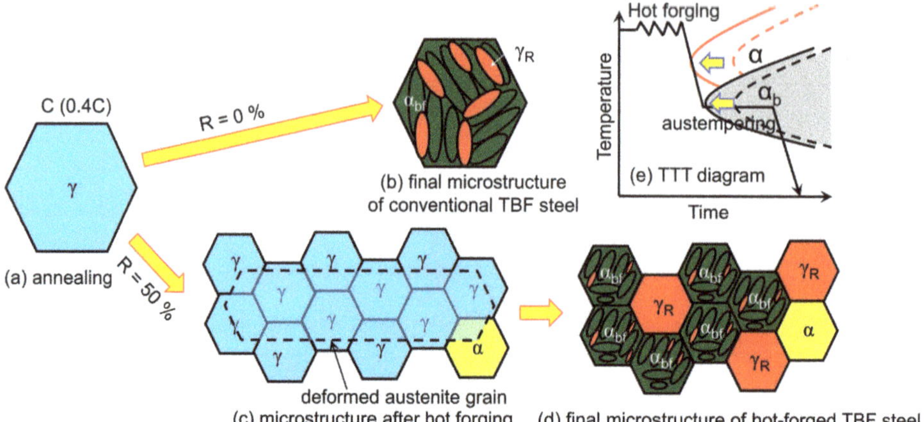



**Figure 12.** Illustrations of microstructure evolution behavior of hot-forged steel C. (**a**) Austenite at annealing temperature, (**b**) final microstructure of conventional TBF steel, (**c**) microstructure after hot forging, (**d**) final microstructure of hot-forged TBF steel, (**e**) TTT diagram, respectively. *R* represents reduction ratio. $\alpha$, $\alpha_b$, $\alpha_{bf}$, $\gamma$, and $\gamma_R$ denote ferrite, bainite, bainitic ferrite, austenite, and retained austenite, respectively. The TTT diagram (**e**) is the time transition temperature diagram.

The yield strength decreased, whereas the tensile strength increased when thermomechanical processing was applied to steels A, B, and C. Moreover, the total elongation decreased, although the uniform elongation hardly changed in steels A, B, and C due to hot forging. As mentioned above, the microstructure of steels A, B, and C possessed fine prior austenitic grain, block, packet, and lath structures, owing to thermomechanical processing. The refinement of the microstructure (i.e., decrease in the grain size) increased the yield strength of the steels according to the Hall–Petch relationship [31,32]. However, it was considered that the yield strength of thermomechanically processed steels A, B, and C decreased because of the formation of ferrite and the premature transformation of block-type retained austenite. In addition, the increase in the tensile strength might have been achieved by the transformation of a large amount of retained austenite at the early stage of plastic deformation, because the volume fraction of the retained austenite increased, and the carbon concentration in the retained austenite decreased, owing to the thermomechanical processing of steels A, B, and C. Moreover, the local deformation capacity might have deteriorated due to the promotion of void initiation at the bainitic ferrite or ferrite matrix and transformed martensite interfaces in the same way as the ferrite-martensite dual-phase steels, because the morphology of the retained austenite varied from filmy to granular, owing to the thermomechanical processing in steels A, B, and C, and the martensite that transformed from block-type retained austenite acted as a hard phase.

### 4.2. Improvement of Hydrogen Embrittlement Properties by Thermomechanical Processing

Steel C showed a decrease in the hydrogen embrittlement susceptibility evaluated by a tensile test and increased the delayed fracture strength evaluated by a four-point bending test, owing to thermomechanical processing, whereas steel A exhibited low hydrogen embrittlement susceptibility and no failure in the four-point bending test, regardless of thermomechanical processing. In general, the hydrogen embrittlement properties of steels A, B, and C with hot forging might be affected by the following factors: variation in the (1) microstructure, (2) characteristics and morphologies of the retained austenite, and (3) hydrogen concentration. For (1) the change in the microstructure owing to thermomechanical processing, it is known that the hydrogen embrittlement properties of high-strength steels are improved by the refinement of the grain size and microstructure.

Zan et al. [33] and Koyama et al. [34] have reported that the hydrogen embrittlement behaviors of the Fe-22Mn-0.6C TWIP steel and the CoCrFeMnNi high-entropy alloy with different grain sizes were investigated, and excellent hydrogen embrittlement resistances were obtained with smaller grain sizes. The grain refinement and the presence of the packet and block as substructures increased the hydrogen embrittlement resistance because of the reduction of the intergranular or quasi-cleavage facet sizes of the fracture surface. Moreover, the refinement of the grain, packet, and block increased the area of the boundaries. When absorbed hydrogen is diffused and trapped at those boundaries, cracking is unlikely to occur because of the decrease in the hydrogen concentration at those boundaries in the steel with fine grains when compared with that with coarse grains. In steels A, B, and C, the refinement of prior austenitic grain, packet, and block sizes was achieved by thermomechanical processing. In particular, steel C possessed a fine bainitic ferrite matrix. Therefore, low hydrogen embrittlement susceptibility and high delayed fracture strength can be achieved by thermomechanical processing. Considering (2) the characteristics and morphologies of retained austenite, it is known that the hydrogen absorption capacity of retained austenite, which is a face-centered cubic (fcc), is higher than that of ferrite with a body-centered cubic (bcc) [35,36]. Thus, hydrogen embrittlement resistance decreased because the cracks were preferentially initiated at the transformed martensite, where hydrogen was supersaturated, and at the bainitic ferrite matrix and transformed martensite interfaces, where hydrogen was diffused and trapped during tensile deformation accompanied with martensitic transformation of the retained austenite. Moreover, it is considered that crack initiation and propagation are accelerated by the premature transformation of retained austenite, which exhibits low stability [37,38]. Thus, the mechanical stability of the retained austenite is considered important for determining the hydrogen embrittlement properties of the steels. It has been reported that the martensitic transformation behavior of retained austenite is affected by the retained austenite morphologies (blocky or filmy types) [39,40] and the surrounding matrix [39–41]. Figure 13 depicts the relationship between the hydrogen embrittlement susceptibility (*HES*) obtained by tensile tests and the retained austenite characteristics in steels A, B, and C. The *HES* tends to increase with an increasing initial volume fraction, initial carbon concentration, and initial total carbon content of the retained austenite. It should be noted that the change in the *HES* might be attributed to the difference in the strength level, resulting from the additive carbon content of the steels. Although the *HES* of steels B and C, which exhibited high volume fractions of retained austenite, were decreased by thermomechanical processing, the effects of the carbon concentration and total carbon content of the retained austenite on the *HES* were small. It was considered that the *HES* of steels B and C with thermomechanical processing was decreased because of the increase in the stability of the retained austenite, owing to its morphology (i.e., fine and film-type retained austenite), the suppression of deformation-induced transformation of the retained austenite, and the corresponding suppressions of crack initiation and propagation, as steels B and C possessed a refined microstructure consisting of fine bainitic ferrite lath structure and fine retained austenite located at the lath boundaries. For (3) the hydrogen concentration, the thermomechanically processed steels A, B, and C possessed a slightly lower hydrogen concentration than conventional steels A, B, and C (Figure 7). It is known that hydrogen absorbed in the steels is trapped at dislocations [42], grain boundaries, lath and packet boundaries [43], and matrix–carbide interfaces [44]. Moreover, austenite of the fcc phase exhibits a high hydrogen absorption capacity when compared with ferrite, bainitic ferrite, and martensite of the bcc phases [35,36]. Figure 14 depicts the relationship between the diffusible hydrogen concentration ($H_C$) and volume fraction of retained austenite ($f_{\gamma 0}$) in steels A, B, and C. The $H_C$ increased with an increase in $f_{\gamma 0}$. However, the $H_C$ of thermomechanically processed steels A, B, and C tended to be lower than those of conventional steels A, B, and C. The increase in the diffusible hydrogen concentration in steels A, B, and C with high additive carbon content is attributed to the increase in the hydrogen trapping sites, such as the retained austenite, which increased with an increasing additive carbon content and the refined bainitic ferrite matrix. On the other hand, it is considered that the decrease in the $H_C$ in steels A, B, and C with thermomechanical processing was attributed to the decrease in the dislocations in the bainitic ferrite matrix due to the promotion of dynamic and/or static recrystallizations during the hot forging and nucleation of ferrite in the matrix, although the hydrogen trapping sites such as refined prior austenitic grain, packet, and block boundaries and the amount of retained austenite increased. Generally, hydrogen embrittlement may occur because hydrogen is accumulated at the prior austenitic grain, packet, and lath boundaries because of the multiplication and movement of dislocations and the stress-assisted diffusion [45,46] of hydrogen during tensile tests with hydrogen. It can be concluded that the hydrogen embrittlement properties of steels B and C were improved by thermomechanical processing, because the hydrogen concentration at crack initiation sites such as the prior austenitic grain, packet, and block boundaries might not have increased, owing to the refinement of the microstructure.

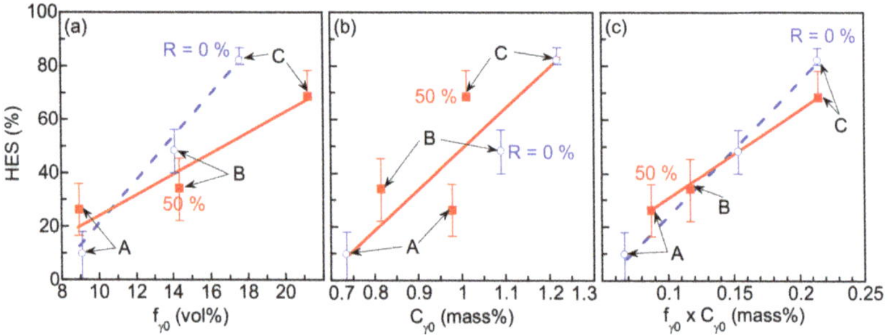


**Figure 13.** Variations in hydrogen embrittlement susceptibility (*HES*) as functions of (**a**) initial volume fraction ($f_{\gamma 0}$), (**b**) initial carbon concentration ($C_{\gamma 0}$), and (**c**) total carbon content ($f_{\gamma 0} \times C_{\gamma 0}$) in retained austenite in steels A, B, and C. *R* represents reduction ratio.

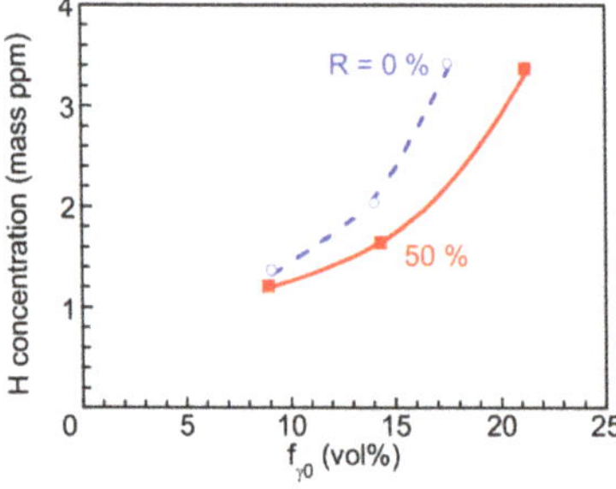


**Figure 14.** Variations in diffusible hydrogen concentration ($H_C$) as a function of initial volume fraction of retained austenite ($f_{\gamma 0}$) in steels A, B, and C. *R* represents reduction ratio.

*4.3. Evaluation of Hydrogen Embrittlement Properties by Tensile Tests and Four-Point Bending Tests*

In this study, the hydrogen embrittlement properties of thermomechanically processed steels A, B, and C were evaluated using the tensile test and four-point bending test methods. In both the tensile and four-point bending tests, it was clarified that the hydrogen embrittlement resistances of steels B and C were improved by thermomechanical processing. In particular, an obvious improvement in the hydrogen embrittlement properties evaluated by the four-point bending tests was obtained, compared with that evaluated by the tensile tests (Figures 6 and 9). As mentioned above, the hydrogen embrittlement resistances of steels A, B, and C might have decreased because a large amount of hydrogen, which was desorbed from transformed martensite, accumulated in the martensite or at the matrix–martensite interfaces, and the crack initiation was accelerated during the tensile tests. Therefore, obvious high hydrogen embrittlement resistances of steels A, B, and C with thermomechanical processing might not have been obtained in the tensile testing results. On the other hand, in the four-point bending tests, the transformation of retained austenite rarely occurred, owing to the constant loading without plastic deformation and the relatively high stability of the retained austenite, although there was a possibility that the retained austenite with low stability transformed to martensite during the four-point bending tests before hydrogen charging. Consequently, although the hydrogen embrittlement resistance evaluated by the tensile test deteriorated, a significant increase in the delayed fracture strength evaluated by the four-point bending test in thermomechanically processed steels B and C could be achieved. It is considered that in service conditions, the plastic deformation and deformation-induced martensitic transformation of retained austenite do not occur because the constant load or cyclic load below the yield strength are applied to the ultrahigh-strength steels in the automobile forging parts. Therefore, it is suggested that the four-point bending test method is more reasonable for accurately evaluating the hydrogen embrittlement properties than the tensile testing method for steels A, B, and C, which undergo martensitic transformation of the retained austenite accompanied by plastic deformation. However, the hydrogen embrittlement resistances of steels B and C with thermomechanical processing improved regardless of the evaluation methods of both the tensile and four-point bending test methods.

## 5. Conclusions

The effects of thermomechanical processing on the hydrogen embrittlement properties of steels A, B, and C were investigated by means of tensile test and four-point bending constant load test to apply the TBF steels to automobile forging parts. The results are summarized as follows:

(1) The microstructures of steels A, B, and C without thermomechanical processing consisted of a bainitic ferrite matrix and retained austenite. On the other hand, the microstructure of steel A was characterized by a fine polygonal ferrite matrix, a bainitic ferrite, and fine retained austenite, whereas the microstructure of steel C consisted of a refined bainitic ferrite matrix with refined prior austenitic grains, packets, blocks, and laths, and blocky and filmy retained austenite when the steels were subjected to thermomechanical processing. These microstructural changes in steels A, B, and C, owing to thermomechanical processing, were attributed to the promotion of the dynamic and static recrystallizations of austenite grains during hot forging after annealing at the austenite region.

(2) In steels A, B, and C without hydrogen, the tensile strength increased from 748 to 837 MPa, from 950 to 959 MPa, and from 1097 to 1122 MPa, respectively, whereas the yield strength decreased from 528 to 500 MPa, from 749 to 559 MPa, and from 937 to 788 MPa, respectively, by thermomechanical processing. On the other hand, the thermomechanical processing decreased the total elongation from 32.6 to 25.4%, from 33.2 to 26.8%, and from 31.9 to 27.6% in steels A, B, and C, respectively, although the uniform elongation hardly changed from 21.4 to 21.5%, from 22.5 to 21.8%, and from 24.4 to 24.1% in steels A, B, and C, respectively.

(3) The hydrogen embrittlement susceptibility evaluated by tensile tests decreased, and the delayed fracture strength evaluated by the four-point bending tests increased because of thermomechanical processing in steels B and C, whereas the hydrogen embrittlement susceptibility slightly increased in hot-forged steel A.

(4) It is considered that the improvements in the hydrogen embrittlement resistances of steels B and C, owing to thermomechanical processing, were attributed to the suppression of the crack initiations at the prior austenitic grain, bainitic ferrite lath, packet, and block boundaries because of the refined microstructure, improvement of the stability of retained austenite, and decrease in the absorbed diffusible hydrogen concentration.

**Author Contributions:** Conceptualization, T.H., J.K., K.-i.S., and A.N.; methodology, T.H., J.K. and K.-i.S.; validation, T.H., M.K., S.A. and E.A.; formal analysis, T.H.; investigation, T.H., Y.Z., J.K., Y.T. and A.N.; resources, T.H., K.-i.S. and E.A.; data curation, T.H.; writing—original draft preparation, T.H.; writing—review and editing, M.K., S.A. and E.A.; visualization, T.H.; supervision, E.A.; project administration, T.H.; funding acquisition, T.H. All authors have read and agreed to the published version of the manuscript.

**Funding:** This work was supported by a Grant-in-Aid for Scientific Research (C), No. JP18K04743. Moreover, part of this work was financially supported by the Advanced Machining Technology & Development Association and The Amada Fundation, No. AF-2009019.

**Institutional Review Board Statement:** Not applicable.

**Informed Consent Statement:** Not applicable.

**Conflicts of Interest:** The authors declare no conflict of interest. The funders had no role in the design of the study; in the collection, analyses, or interpretation of data; in the writing of the manuscript; or in the decision to publish the results.

# References


1. Matlock, D.K.; Speer, J.G.; De Moor, E.; Gibbs, P.J. Recent developments in advanced high strength sheet steels for automotive applications: An overview. *JESTECH* **2012**, *15*, 1–12.
2. Heibel, S.; Dettinger, T.; Nester, W.; Clausmeyer, T.; Tekkaya, A.E. Damage mechanisms and mechanical properties of high-strength multiphase steels. *Materials* **2018**, *11*, 761. [CrossRef] [PubMed]
3. Kozłowska, A.; Grzegorczyk, B.; Morawiec, M.; Grajcar, A. Explanation of the PLC effect in advanced high-strength medium-Mn steels. A review. *Materials* **2019**, *12*, 4175. [CrossRef] [PubMed]
4. Zackay, V.F.; Parker, E.R.; Fahr, D.; Bush, R. The enhancement of ductility in high-strength steels. *Trans. Am. Soc. Met.* **1967**, *60*, 252–259.
5. Branco, R.; Berto, F. Mechanical behavior of high-strength, low-alloy steels. *Metals* **2018**, *8*, 610. [CrossRef]
6. Sugimoto, K.; Tsunezawa, M.; Hojo, T.; Ikeda, S. Ductility of 0.1-0.6C-1.5Si-1.5Mn ultra high-strength TRIP-aided sheet steels with bainitic ferrite matrix. *ISIJ Int.* **2004**, *44*, 1608–1614. [CrossRef]
7. Sugimoto, K.; Hojo, T.; Srivastava, A.K. An overview of fatigue strength of case-hardening TRIP-aided martensitic steels. *Metals* **2018**, *8*, 355. [CrossRef]
8. Sugimoto, K.; Mizuno, Y.; Natori, M.; Hojo, T. Effects of fine particle peening on fatigue strength of a TRIP-aided martensitic steel. *Int. J. Fatigue* **2017**, *100*, 206–214. [CrossRef]
9. Hojo, T.; Kobayashi, J.; Sugimoto, K. Impact properties of low-alloy transformation-induced plasticity-steels with different matrix. *Mater. Sci. Technol.* **2016**, *32*, 1035–1042. [CrossRef]
10. Caballero, F.G.; Roelofs, H.; Hasler, S.; Capdevila, C.; Chao, J.; Cornide, J.; Garcia-Mateo, C. Influence of bainite morphology on impact toughness of continuously cooled cementite free bainitic steels. *Mater. Sci. Technol.* **2013**, *28*, 95–102. [CrossRef]
11. Kvackaj, T.; Bidulská, J.; Bidulský, R. Overview of HSS steel grades development and study of reheating condition fffects on austenite grain size changes. *Materials* **2021**, *14*, 1988. [CrossRef]
12. Prislupčák, P.; Kvačkaj, T.; Bidulská, J.; Záhumenský, P.; Homolová, V.; Zimovčák, P. Austenite-ferrite transformation temperatures of C Mn Al HSLA steel. *Acta Metall. Slovaca* **2021**, *27*, 207–209. [CrossRef]
13. Zhao, J.; Jiang, Z. Thermomechanical processing of advanced high strength steels. *Prog. Mater. Sci.* **2018**, *94*, 174–242. [CrossRef]
14. Sugimoto, K.; Hayakawa, A.; Hojo, T.; Hashimoto, S.; Ikeda, S. Grain refinement of high strength low alloy TRIP-aided ferrous steels by thermomechanical processing in alpha plus gamma region. *Tetsu Hagane* **2003**, *89*, 1233–1239. [CrossRef]
15. Sugimoto, K.; Itoh, M.; Hojo, T.; Hashimoto, S.; Ikeda, S.; Arai, G. Microstructure and mechanical properties of ausformed ultra high-strength TRIP-aided steels. *Mater. Sci. Forum* **2007**, *539-543*, 4309–4314. [CrossRef]
16. Sugimoto, K.; Hojo, T.; Srivastava, A. Low and medium carbon advanced high-strength forging steels for automotive applications. *Metals* **2019**, *9*, 1263. [CrossRef]
17. Sugimoto, K.; Sato, S.; Kobayashi, J.; Srivastava, A.K. Effects of Cr and Mo on mechanical properties of hot-forged medium carbon TRIP-aided bainitic ferrite steels. *Metals* **2019**, *9*, 1066. [CrossRef]
18. Hojo, T.; Kobayashi, J.; Kochi, T.; Sugimoto, K. Effects of thermomechanical processing on microstructure and shear properties of 22SiMnCrMoB TRIP-aided martensitic steel. *Iron Steel Technol.* **2015**, *10*, 102–110.
19. Hojo, T.; Kochi, T.; Sugimoto, K. Effects of warm working on microstructural and shear deformation properties of TRIP-aided martenitic steel. *Mater. Sci. Forum* **2017**, *879*, 2312–2317. [CrossRef]
20. Li, X.; Ma, X.; Zhang, J.; Akiyama, E.; Wang, Y.; Song, X. Review of hydrogen embrittlement in metals: Hydrogen diffusion, hydrogen characterization, hydrogen embrittlement mechanism and prevention. *Acta Metall. Sin. (Engl. Lett.)* **2020**, *33*, 759–773. [CrossRef]
21. Martínez, C.; Briones, F.; Villarroel, M.; Vera, R. Effect of atmospheric corrosion on the mechanical properties of SAE 1020 structural steel. *Materials* **2018**, *11*, 596. [CrossRef] [PubMed]

22. Sojka, J.; Vodárek, V.; Schindler, I.; Ly, C.; Jérôme, M.; Váňová, P.; Ruscassier, N.; Wenglorzová, A. Effect of hydrogen on the properties and fracture characteristics of TRIP 800 steels. *Corros. Sci.* **2011**, *53*, 2575–2581. [CrossRef]
23. Laureys, A.; Depover, T.; Petrov, R.; Verbeken, K. Characterization of hydrogen induced cracking in TRIP-assisted steels. *Int. J. Hydrog. Energy* **2015**, *40*, 16901–16912. [CrossRef]
24. Hojo, T.; Kikuchi, R.; Waki, H.; Nishimura, F.; Ukai, Y.; Akiyama, E. Effect of strain rate on the hydrogen embrittlement property of ultra high-strength low alloy TRIP-aided steel. *ISIJ Int.* **2018**, *58*, 751–759. [CrossRef]
25. Hojo, T.; Kumai, B.; Koyama, M.; Akiyama, E.; Waki, H.; Saitoh, H.; Shiro, A.; Yasuda, R.; Shobu, T.; Nagasaka, A. Hydrogen embrittlement resistance of pre-strained ultra-high-strength low alloy TRIP-aided steel. *Int J. Fract.* **2020**, *224*, 253–260. [CrossRef]
26. Hojo, T.; Kobayashi, J.; Sugimoto, K.; Nagasaka, A.; Akiyama, E. Effects of alloying elements addition on delayed fracture properties of ultra high-strength TRIP-aided martensitic steels. *Metals* **2019**, *10*, 6. [CrossRef]
27. Dyson, D.J.; Holmes, B. Effect of alloying additions on the lattice parameter of austenite. *J. Iron Steel Inst.* **1970**, *208*, 469–474.
28. Hojo, T.; Sugimoto, K.; Mukai, Y.; Ikeda, S. Effects of aluminum on delayed fracture properties of ultra high strength low alloy TRIP-aided steels. *ISIJ Int.* **2008**, *48*, 824–829. [CrossRef]
29. Silveira, A.C.D.F.; Bevilaqua, W.L.; Dias, V.W.; De Castro, P.J.; Epp, J.; Rocha, A.D.S. Influence of hot forging parameters on a low carbon continuous cooling bainitic steel microstructure. *Metals* **2020**, *10*, 601. [CrossRef]
30. Nürnberger, F.; Grydin, O.; Schaper, M.; Bach, F.W.; Koczurkiewicz, B.; Milenin, A. Microstructure transformations in tempering steels during continuous cooling from hot forging temperatures. *Steel Res. Int.* **2010**, *81*, 224–233. [CrossRef]
31. Takaki, S.; Kawasaki, K.; Kimura, Y. Mechanical properties of ultra fine grained steels. *J. Mater. Process. Technol.* **2001**, *117*, 359–363. [CrossRef]
32. Bergström, Y.; Hallén, H. Hall–petch relationships of iron and steel. *Met. Sci.* **2013**, *17*, 341–347. [CrossRef]
33. Zan, N.; Ding, H.; Guo, X.; Tang, Z.; Bleck, W. Effects of grain size on hydrogen embrittlement in a Fe-22Mn-0.6C TWIP steel. *Int. J. Hydrog. Energy* **2015**, *40*, 10687–10696. [CrossRef]
34. Koyama, M.; Ichii, K.; Tsuzaki, K. Grain refinement effect on hydrogen embrittlement resistance of an equiatomic cocrfemnni high-entropy alloy. *Int. J. Hydrog. Energy* **2019**, *44*, 17163–17167. [CrossRef]
35. Hojo, T.; Koyama, M.; Terao, N.; Tsuzaki, K.; Akiyama, E. Transformation-assisted hydrogen desorption during deformation in steels: Examples of α′- and ε-martensite. *Int. J. Hydrog. Energy* **2019**, *44*, 30472–30477. [CrossRef]
36. Koyama, M.; Yamasaki, D.; Ikeda, A.; Hojo, T.; Akiyama, E.; Takai, K.; Tsuzaki, K. Detection of hydrogen effusion before, during, and after martensitic transformation: Example of multiphase transformation-induced plasticity steel. *Int. J. Hydrog. Energy* **2019**, *44*, 26028–26035. [CrossRef]
37. Timokhina, I.B.; Hodgson, P.D.; Pereloma, E.V. Effect of microstructure on the stability of retained austenite in transformation-induced-plasticity steels. *Metall. Mater. Trans. A* **2004**, *35*, 2331–2341. [CrossRef]
38. Zwaag, S.V.D.; Zhao, L.; Kruijver, S.O.; Sietsma, J. Thermal and mechanical stability of retained austenite in aluminum-containing multiphase TRIP steels. *ISIJ Int.* **2004**, *42*, 1565–1570. [CrossRef]
39. Xiong, X.C.; Chen, B.; Huang, M.X.; Wang, J.F.; Wang, L. The effect of morphology on the stability of retained austenite in a quenched and partitioned steel. *Scr. Mater.* **2013**, *68*, 321–324. [CrossRef]
40. Shen, Y.F.; Qiu, L.N.; Sun, X.; Zuo, L.; Liaw, P.K.; Raabe, D. Effects of retained austenite volume fraction, morphology, and carbon content on strength and ductility of nanostructured TRIP-assisted steels. *Mater. Sci. Eng. A* **2015**, *636*, 551–564. [CrossRef]
41. Jacques, P.J.; Ladrie'Re, J.; Delannay, F. On the influence of interactions between phases on the mechanical stability of retained austenite in transformation-induced plasticity multiphase steels. *Metall. Mater. Trans. A* **2001**, *32*, 2759–2768. [CrossRef]
42. Bhadeshia, H.K.D.H. Prevention of hydrogen embrittlement in steels. *ISIJ Int.* **2016**, *56*, 24–36. [CrossRef]
43. Momotani, Y.; Shibata, A.; Terada, D.; Tsuji, N. Effect of strain rate on hydrogen embrittlement in low-carbon martensitic steel. *Int. J. Hydrog. Energy* **2017**, *42*, 3371–3379. [CrossRef]
44. Nagao, A.; Hayashi, K.; Oi, K.; Mitao, S. Effect of uniform distribution of fine cementite on hydrogen embrittlement of low carbon martensitic steel plates. *ISIJ Int.* **2012**, *52*, 213–221. [CrossRef]
45. Takagi, S.; Hagihara, Y.; Hojo, T.; Urushihara, W.; Kawasaki, K. Comparison of hydrogen embrittlement resistance of high strength steel sheets evaluated by several methods. *ISIJ Int.* **2016**, *56*, 685–692. [CrossRef]
46. Hirth, J.P.; Carnahan, B. Hydrogen adsorption at dislocations and cracks in Fe. *Acta Metall.* **1978**, *26*, 1795–1803. [CrossRef]

Article

# Air-Hardening Die-Forged Con-Rods—Achievable Mechanical Properties of Bainitic and Martensitic Concepts

Alexander Gramlich [1,*], Robert Lange [2], Udo Zitz [3] and Klaus Büßenschütt [1]

[1]  Steel Institute (IEHK) of RWTH Aachen University, Intzestraße 1, 52072 Aachen, Germany; Klaus.Buessenschuett@iehk.rwth-aachen.de

[2]  Lech-Stahlwerke GmbH, Industriestraße 1, 86405 Meitingen, Germany; robert.lange@lech-stahlwerke.de

[3]  Frauenthal Powertrain Management GmbH & Co. KG, In den Hofwiesen 13, 58840 Plettenberg, Germany; udo.zitz@fta-group.com

*  Correspondence: alexander.gramlich@iehk.rwth-aachen.de; Tel.: +49-241-80-95786

**Abstract:** Three air-hardening forging steels are presented, concerning their microstructure and their mechanical properties. The materials have been produced industrially and achieve either bainitic or martensitic microstructures by air-cooling directly from the forging heat. The bainitic steels are rather conservative steel concepts with an overall alloy concentration of approximately 3 wt.%, while the martensitic concept is alloyed with 4 wt.% manganese (and additional elements), and therefore belongs to the recently developed steel class of medium manganese steels. The presented materials achieve high strengths (YS: 720 MPa to 850 MPa, UTS: 1055 MPa to 1350 MPa), good elongations ($A_u$: 4.0% to 5.9%, $A_t$: 12.3% to 14.9%), and impact toughnesses (up to 37 J) in the air-hardened condition. It is shown that air-hardened steels achieve properties close to standard Q + T steels, while being produced with a significantly reduced heat treatment.

**Keywords:** high-strength steels; forging; air-cooling; mechanical properties; microstructure

**Citation:** Gramlich, A.; Lange, R.; Zitz, U.; Büßenschütt, K. Air-Hardening Die-Forged Con-Rods—Achievable Mechanical Properties of Bainitic and Martensitic Concepts. *Metals* **2022**, *12*, 97. https://doi.org/10.3390/met12010097

Academic Editor: Mohammad Jahazi

Received: 1 December 2021
Accepted: 31 December 2021
Published: 4 January 2022

**Publisher's Note:** MDPI stays neutral with regard to jurisdictional claims in published maps and institutional affiliations.

## 1. Introduction

Quenching and tempering steels (Q + T) are widely used in the forging industry if the produced component requires high strength as well as high toughness. This property balance is achieved by a three-stepped heat treatment of austenization, quenching, and tempering, which transforms the brittle as-quenched martensite into a technical usable tempered material, but the long heat treatment procedure introduces high costs and $CO_2$ emissions. Additionally, the quenching results in distortion of the forged components which can lead to extensive machining adding additional cost (and $CO_2$-emissions) to the final product. To address this issue, different steel concepts have been developed in the past, trying to achieve the properties of Q + T steels by air-cooling from the forging heat [1]. Precipitation-hardening ferritic pearlitic steels (PHFP) were the first attempt to achieve this goal and are already widely used in the industry, but the properties of Q + T steels stay unreached [1]. The PHFP-steels were succeeded by different grades of bainitic steels, which achieve higher ductilities and impact toughness by strain-induced martensite formation of retained austenite [2]. Additions of molybdenum and chromium have been succesfully used to further increase the impact toughness and the strength of these materials [3]. The latter even increased the resistance against hydrogen embrittlement [4]; however, these concepts are limited to components with thin wall thicknesses, as the temperature must be controlled precisely to obtain the necessary complex microstructure of bainite and retained austenite. Several concepts for hot forging have been developed based on the material group of advanced high-strength steels (AHSS) of the third generation. This concepts contain isothermal transformation processes after hot forging, to improve the mechanical properties [5].

Besides the PHFP-steels, different steel concepts have been developed [1] which form bainitic [2,6] or martensitic microstructures [7] after air-cooling. The use of bainitic microstructures is limited to components with small wall thicknesses. The reason for this is the different local cooling paths of large components (difference between edge and core) [2,6], which can lead to different microstructures and thus to a gradient in the mechanical properties. In particular, reheating due to bainitic phase transformation affects the local time–temperature profiles. Therefore, air-hardening martensitic steels are more promising for forging components with large cross sections, as the martensitic transformation is an athermal process and the process window for cooling with similar mechanical properties is wider. These air-hardening concepts show static and cyclic properties comparable to, or even higher than, Q + T steels, which has already been proven on a laboratory [8] and an industrial scale [7,9]. However, a broad application is prevented so far by a lack of available data of industrial trials.

Con-rods, used in engines of different kinds, are standard die-forging products, as they combine the need for a high-strength steel and are produced in large quantities. Con-rods were originally produced from Q + T steels but, due to the abovementionend reasons, the focus shifted to PHFP steels [10]. Due to the direct influence of the con-rods on the fuel consumption, different studies on the lightweighting potential [11,12] or the general performance [13,14] were carried out in the past. Besides die-forging, sinter die forging of con-rods has been investigated as an alternative process. However, it was demonstrated that the die-forged steels have a much higher fatigue resistance than the sinter components, which prohibited the broad application of this elaborate process [13]. As con-rods are produced by fracture-splitting [15–17], the fracture behaviour of steels for this application needs to be investigated. Contrary to other applications, the material needs to split in a brittle manner, as ductile deformation might lead to quality issues [17,18] or completely prevent the splitting process.

The aim of this study is to investigate the usability of recently developed air-hardening steels for forged con-rod application. For this matter, two optimized bainitic steel grades (35MnCrB6-4 and 19MnCrMo7-6, consecutively called B1 and B2) and a recently developed martensitic concept (consecutively called M) were industrially produced and forged into semifinished con-rods, which have been characterized concerning their microstructure and their mechanical properties. It is demonstrated that these steels reach mechanical properties which fit the requirements of specific applications, in this case con-rods, while being produced with a reduced heat treatment in comparison to standard quench and tempering steels. Contrary to the majority of studies available, the presented results are obtained from industrial manufactured components. Therefore, this study demonstrates the efforts of the forging industry to reduce the $CO_2$-equivalent of steel products.

## 2. Materials and Methods

The investigated materials were industrially produced at Lech-Stahlwerke GmbH (alloys 35MnCrB6-4 'B1' and 19MnCrMo7-6 'B2', commercially available steel grades) and at BGH Edelstahl Siegen GmbH (alloy 15MnSi16-2 "M", produced during the research project "IGF 27 EWN" [9]). The chemical compositions are displayed in Table 1. The steels B1 and B2 were melted in an electric arc furnace (EAF) with a charge weight of 85 t. After a vacuum treatment, the steels were continuously casted in a bow-type continuous caster with a casting format of $240 \times 240 \, mm^2$. Steel M1 was melted in an EAF as well, but consecutive to the vacuum treatment casted via ingot casting. The charge weight was 50 t, with each ingot weighing 3.3 t. The ingots had a conical casting format with a diameter from 420 mm to 520 mm. After the casting, all steel grades were hot-rolled to a format of $130 \times 130 \, mm^2$ with a rolling temperature between 1220 and 1250 °C. Prior to the forging, all steels were heated to 1250 °C. Die forging was performed approximately 15 s after the reheating, with a material surface temperature of 1200 °C. After the forging, the components were air-cooled to room temperature on a conveyer belt. No artificial air flow was created; the components were cooled in resting air. The forging and the consecutive heat treatment was performed at Frauenthal Powertrain GmbH & Co. KG. The semifinished products have a thickness between 22 mm to 28 mm, a length of 220 mm, and a width between 40 mm to 80 mm, as displayed in Figure 1. The martensitic components (produced from steel M) were additionally batch-annealed at 450 °C for 2 h (this state is consecutively called "M + ba"). The samples for microstructure investigation were mechanically grinded and polished with diamond slurry. Consecutively, the samples were etched with either Nital etchant for general analyses or Picric etchant for the determination of the prior austenite grain size. The prior austenite grain size was determined by the line intercept method. For each steel, at least 300 grains were measured. Scanning electron microscopy (SEM) experiments were performed on aZeiss Sigma SEM, using multiple detectors and parameters. Tensile tests were performed on cylindric specimens with a diameter of 5 mm and a gauge length of 25 mm with a constant strain rate of $0.008\,s^{-1}$ using an ZwickRoell Z100. Instrumented Charpy V-notch samples were tested at room temperature using a 300 J hammer, produced by Losenhausenwerk AG. Vickers hardness measurements [19] were performed on an Instron Wolpert Dia Testor using a testing force of 294.2 N (HV30).

**Table 1.** Chemical compositions of the investigated steels determined by spark spectral analysis. The steels of type 35MnCrB6-4 (modified 33MnCrB5-2/1.7185), 19MnCrMo7-6 (1.7971) and 15MnSi16-2 (1.5132) are abbreviated to "B1", "B2", and "M", respectively. All concentrations are given in wt.%.

| Alloy | C * | Si | Mn | P | S * | Cr | Mo | Al | Nb | B | N |
|---|---|---|---|---|---|---|---|---|---|---|---|
| B1 | 0.34 | 0.19 | 1.61 | 0.009 | 0.006 | 0.80 | max. 0.30 | 0.027 | 0.002 | 0.0036 | 0.007 |
| B2 | 0.20 | 0.26 | min. 1.50 max. 2.10 | 0.010 | 0.007 | min. 1.40 max. 1.85 | max. 0.40 | 0.023 | max. 0.060 | max. 0.003 | 0.019 |
| M | 0.15 | 0.50 | 3.90 | 0.004 | 0.002 | 0.10 | 0.24 | 0.52 | 0.030 | 0.0025 | 0.006 |

* C, S determined with LECO combustion analysis.

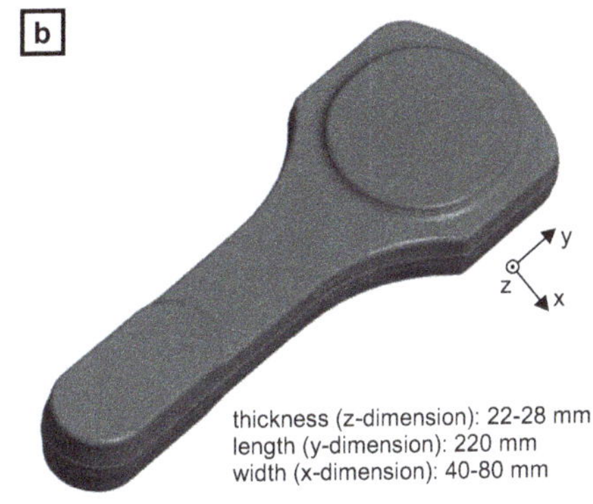


**Figure 1.** Shape and geometry of the semifinished con-rod. (**a**) Picture of the components after forging at a conveyor belt; (**b**) computer model of the component.

## 3. Results

### 3.1. Microstructure

Steels B1 and B2 show a thoroughly bainitic structure, as displayed in Figure 2. Specimen B1 shows a coarse structure of the bainite constituents (bainite grains, lamelas) with a grain size of 10 μm to 30 μm, which consist mainly of lamellar bainite, with large, globular-shaped bainite grains in between. The structure of B2 is finer and the global bainite grains are, with around 10 μm in size, much smaller. The globular bainite shows an elongated morphology and is evenly distributed between the lancets. The specimens M and M + ba show a typical martensitic microstructure with lancets and a similar grain size. Adjacent to the fine martensitic structure, small carbide accumulations, with a size under 10 μm, were found. Smaller magnifications of different regions of the component reveals lamellar superstructures in rolling direction for steel B1, as seen in Figure 3a,b. The structures appear globular when they are observed parallel to the rolling direction, as seen in Figure 3b,c. The other materials do not show these deformation structures.

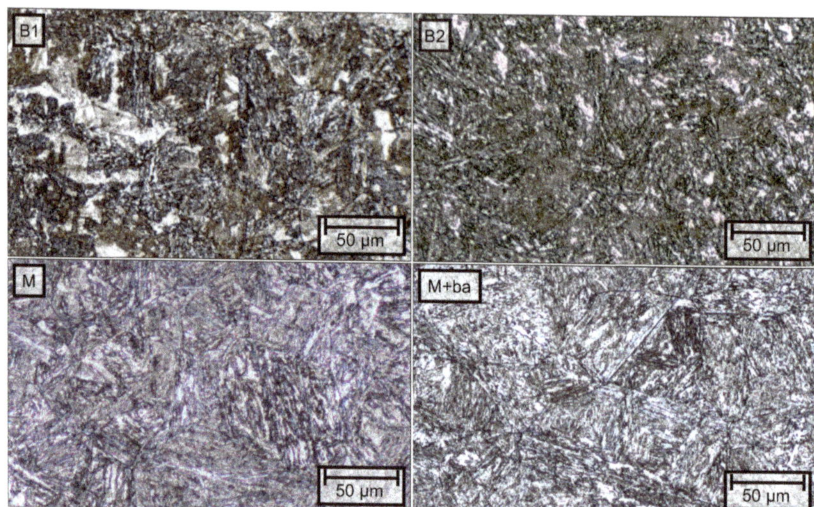

**Figure 2.** Microstructure of the bainitic steel grades (B1 and B2) as well as the martensitic steel in the as-forged condition (M) and in the annealed condition (M + ba).

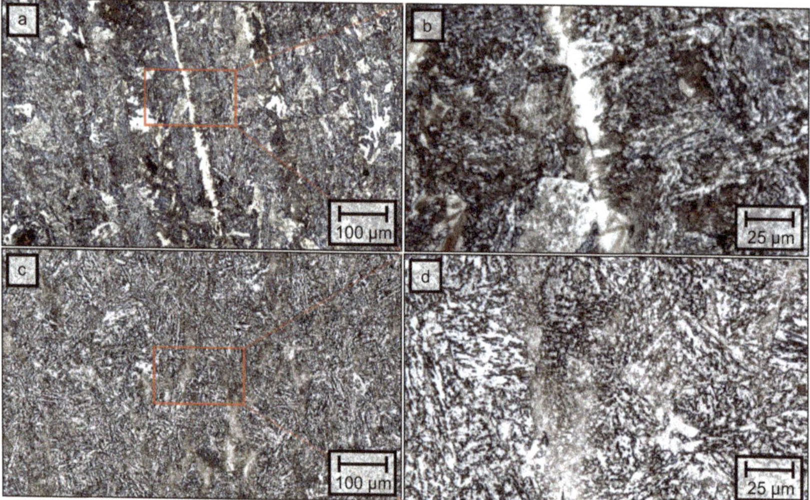

**Figure 3.** Microstructural inhomogeneities (superstructures) visible in the bainitic materials. The superstructures appear elongated when they are examined parallel to the rolling direction (**a**,**b**) and globular when examined vertical to the rolling direction (**c**,**d**).

The local size of the bainite constituents in the center of the component is, with 10 μm to 30 μm, slightly smaller than on the edges, with 30 μm to 50 μm. The same applies to the local size distribution of material B2. M and M + ba show no size alteration of the martensite laths between the edges and the inner area of the component. The accumulated carbides, however, are larger at the edges of the specimen than in the core and build longitudinal structures between the martensite lancets, with a similar size and direction.

Taking a look at the precipitations, B1 shows titanium nitrides with a size under 5 µm (Figure 4a). They can be found sporadically, but are not representative for the alloy. In M and M + ba, many carbides can be found, which accumulate in stretched areas between the martensite lancets. After the annealing, these carbides have grown in size forming larger carbide areas, as shown in Figure 4b). These areas reach sizes of 10 µm to 15 µm. Additionally, the M and M + ba materials show small, globular manganese sulfides, with a diameter of under 5 µm.

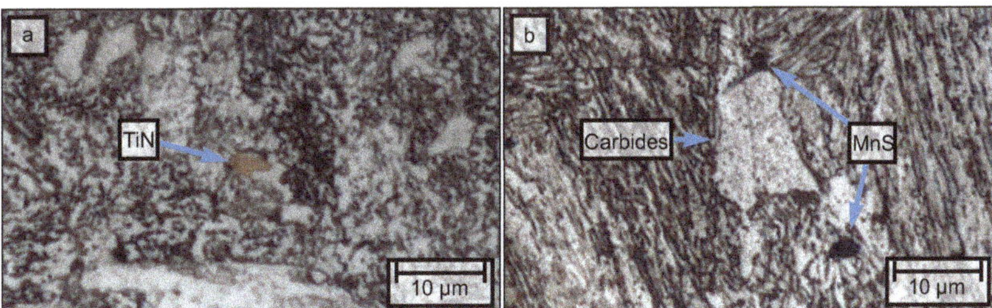


**Figure 4.** Precipitates observed by optical microscopy: (**a**) titanium nitride, (**b**) carbide agglomeration and manganese sulfides.

The carbides in the martensitic steel are barely observable by LOM. SEM images reveal (Figure 5) that the carbides have a length between 100 nm to 300 nm and a thickness smaller than 50 nm.

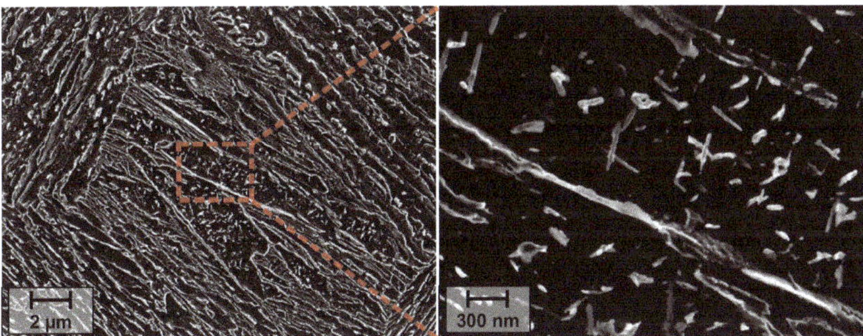


**Figure 5.** Carbide structures observed by SEM in the "M" steel after air-hardening.

The prior austenite grain size of the materials was measured by the line intercept method. For each material, a dataset of approximately 300 measurements was collected. To reveal the prior austenite grain size, picric etching was performed on the steels. The resulting microstructures with visualized prior austenite grain boundaries are displayed in Figure 6. All three steels show globular prior austenitic grains, while B1 and B2 are smaller than M. The dark areas on the specimen are caused by etching artefacts. The arithmetic mean of the prior austenite grain size for the steels B1, B2, and M were measured to be 34 µm, 34 µm, and 55 µm, respectively. In order to characterize the distribution of the measured grain sizes, box plot diagrams and histograms are displayed in Figure 7.

The box plot diagram reveals that the interquartile range of the bainitic alloys is approximately from 20 µm to 50 µm while it is from 35 µm to 75 µm for the martensitic grade. While the largest grains were measured to be slightly below 100 µm for the bainitic grades, steel M showed some outliers with a prior austenite grain size of up to 150 µm. The medians of the dataset were all found to be slightly lower than the arithmetic mean, with a maximum difference for steel M of 5 µm. An analysis of the histograms reveals comparable observations. The arithmetic mean (visualized by the dashed line) lies on the left of the peak of the Weibull fit. If the bainitic steels are compared with the martensitic one, it can be seen that the width of the Weibull fit is increased for the martensitic steel, while the height is reduced. The bainitic steels have a similar shape, with a steep climb to 35 µm, followed by a hyperbolic fall to 110 µm. The grain size distribution of M is shifted to larger grains and appears more inhomogeneous, overall. The Weibull plot climbs less steeply to the maximum of 40 µm and falls hyperbolically to a grain size of 160 µm.

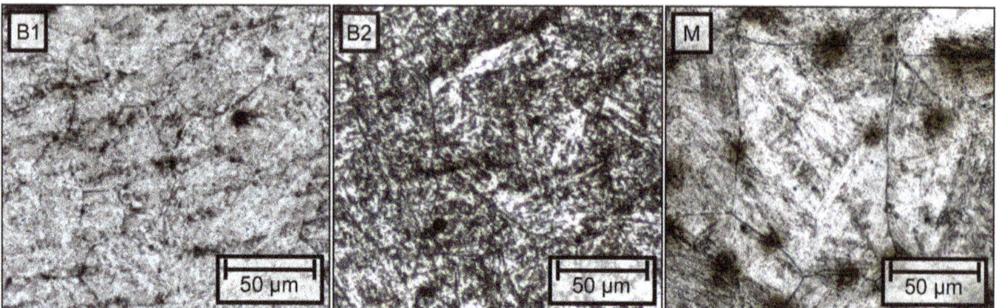


**Figure 6.** Prior austenite grains of the investigated bainitic (B1 and B2) and martensitic (M) steel, revealed by picric etching.

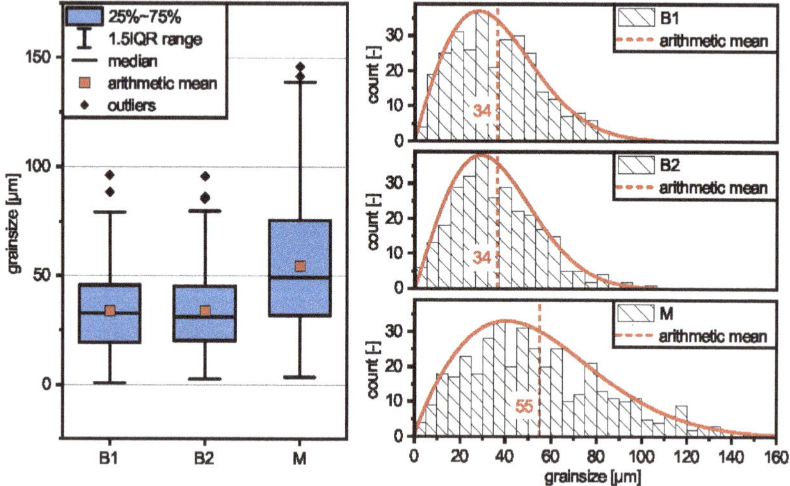


**Figure 7.** Visualization of the grain size distribution of the investigated materials determined by line intercept method. The box diagram on the left displays the medians and arithmetic means, interquartile ranges (IQR), and outliers of the dataset. Additionally, the grain size histogram is displayed on the right together with the corresponding Weibull fit and the arithmetic mean, which is represented by the dashed line.

### 3.2. Fracture Surfaces

In addition to the microstructure, fracture surfaces after Charpy V-notch impact test were observed using SE imaging and EDX analyses. A comparison of the overall fracture morphology is displayed in Figure 8. All samples are oriented with the notch positioned on the left edge of the image. The bainitic steel grades B1 and B2 show a complete brittle fracture, while the martensitic grades show regions which failed by ductile fracture. These regions are on the edge of the sample and have a thickness of approximately 1000 µm, resulting in an area fraction of ductile fracture of approximately 25%. The brittle areas of all investigated samples show a mixture of intra- and transgranular fracture at higher magnifications, as displayed in Figure 9. The bainitic grade B1 shows facets with a diameter of >30 µm, while the other materials have smaller facet diameters. Additionally, the amount of intragranular fractures seems to be higher for the bainitic grades.

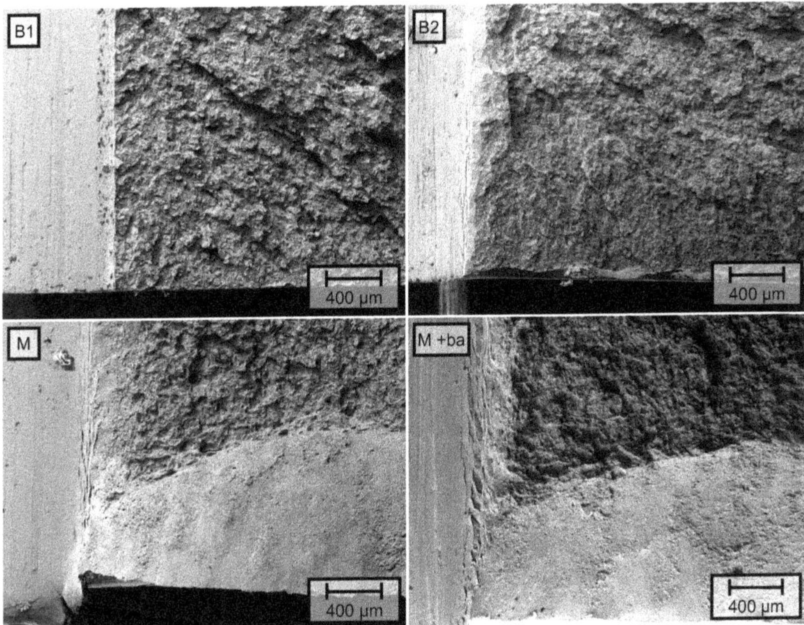

**Figure 8.** Fracture surface of Charpy V-notch impact test samples: Overview of the different overall fracture morphology, with the notch visible on the left edge of the image.

**Figure 9.** Fracture surface of Charpy V-notch impact test samples: Comparison of the areas where mostly brittle fracture can be seen.

At even higher magnifications (Figure 10), it can be seen that the martensitic grades (M and M + ba) have ductile regions on the edges of the facets. Figure 10 further reveals different precipitates or inclusions at the bottom of pores which were investigated by EDX measurements. The results of the EDX measurements 1 to 7 are displayed in Table 2. Measurements 1, 2, 3, 4, and 7 mainly show large increases in the manganese and sulfur concentration while measurements 5 and 6 show larger increases in the aluminum and nitrogen concentration. Scan 3 shows, besides the increases in manganese and sulfur, additional increases in aluminum and nitrogen.

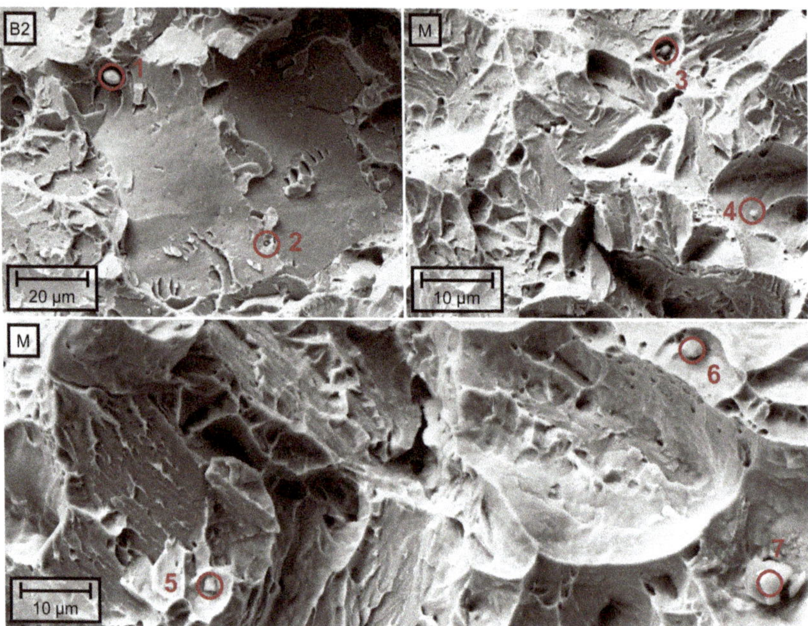


**Figure 10.** Fracture surface of Charpy V-notch impact test samples: Observed particles which were investigated with EDX, consecutively.

**Table 2.** Chemical composition determined with EDX measurements. All concentrations are given in wt.%.

| Scan-# | Fe | | Mn | | S | | Al | | Cr | | N | | Mg | | Ca | |
|---|---|---|---|---|---|---|---|---|---|---|---|---|---|---|---|---|
| | Signal | σ | Signal | σ | Signal | σ | Signal | σ | Signal | σ | Signal | σ | Signal | σ | Signal | σ |
| 1 | 8.5 | 0.19 | 51.9 | 0.43 | 28.6 | 0.26 | - | - | 0.6 | 0.07 | - | - | - | - | 0.9 | 0.05 |
| 2 | 45.1 | 0.32 | 29.1 | 0.25 | 15.8 | 0.15 | - | - | 1.2 | 0.07 | - | - | - | - | - | - |
| 3 | 23.6 | 0.30 | 32.3 | 0.36 | 20.7 | 0.23 | 4.3 | 0.11 | - | - | 4.4 | 0.53 | 2.1 | - | 0.6 | 0.05 |
| 4 | 55.7 | 0.44 | 5.1 | 0.13 | 10.3 | 0.13 | - | - | - | - | - | - | 0.6 | 0.09 | 10.5 | 0.13 |
| 5 | 25.5 | 0.27 | 3.3 | 0.10 | 2.0 | 0.06 | 32.3 | 0.30 | - | - | 29.3 | 0.49 | 0.4 | 0.05 | 0.5 | 0.04 |
| 6 | 14.1 | 0.20 | 1.2 | 0.07 | 0.8 | 0.04 | 43.7 | 0.43 | - | - | 32.3 | 0.52 | - | - | 0.5 | 0.04 |
| 7 | 10.6 | 0.20 | 21.7 | 0.26 | 31.2 | 0.31 | 1.5 | 0.09 | - | - | - | - | 7.0 | 0.13 | 15.0 | 0.18 |
| 8 | 86.3 | 0.33 | 4.3 | 0.12 | - | - | 0.4 | 0.06 | 0.2 | 0.05 | - | - | - | - | - | - |
| 9 | 4.4 | 0.15 | 33.5 | 0.33 | 25.3 | 0.25 | 1.5 | 0.07 | - | - | - | - | 5.1 | 0.10 | 3.1 | 0.07 |
| 10 | 46.3 | 0.32 | 3.3 | 0.09 | 0.6 | 0.04 | 19.9 | 0.17 | - | - | - | - | - | - | - | - |
| 11 | 24.4 | 0.26 | 6.8 | 0.12 | 5.2 | 0.08 | 21.3 | 0.21 | - | - | 18.6 | 0.51 | 0.8 | 0.04 | 0.6 | 0.04 |
| 12 | 4.4 | 0.14 | 31.9 | 0.31 | 23.5 | 0.23 | 1.3 | 0.06 | - | - | - | - | 4.7 | 0.09 | 2.8 | 0.06 |
| 13 | 34.6 | 0.33 | 14.2 | 0.18 | 8.5 | 0.11 | 10.2 | 0.13 | - | - | 12.1 | 0.49 | 1.5 | 0.06 | 1.1 | 0.05 |

Precipitates with similar compositions were found during the characterization of the microstructure, as exemplary shown in Figure 11. The EDX measurements (Table 2) reveal a manganese-rich and an aluminum-rich particle (scan 9 and 10, respectively). Higher magnification of the manganese-rich particle reveals areas of different contrast very close to the globular main particle. An EDX mapping (Figure 11) shows that these areas are enriched with aluminum. Additionally, the mapping demonstrates the enrichment of the further alloying elements in the precipitate, namely magnesium and calcium.

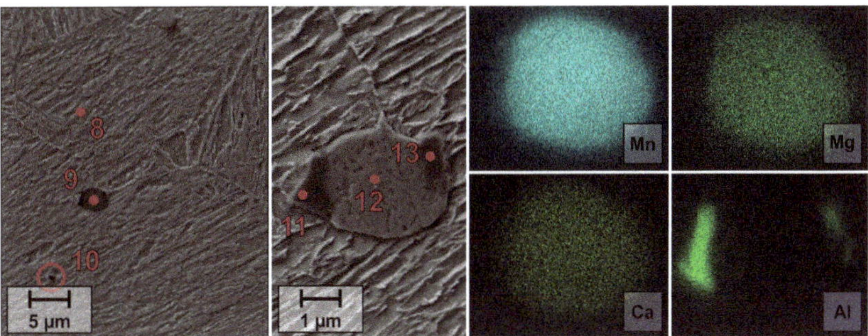


**Figure 11.** Microstructure of steel "M" observed by SEM. Different types of precipitates are displayed and the positions of EDX analyses are indicated. The figure is completed by an EDX mapping of aluminum and manganese, as well as the trace elements magnesium and calcium.

### 3.3. Mechanical Properties

The mechanical properties of the alloys were investigated by tensile tests, instrumented Charpy V-notch tests, and Vickers hardness measurements. The obtained values are summarized in Table 3, and are augmented by the properties of the standard quenched and tempered steel 42CrMo4 (AISI 4140) [7]. The bainitic materials do not differ significantly, concerning their strength. Alloys B1 and B2 reach yield strengths (YS) of 732 MPa and 721 MPa, as well as ultimate tensile strengths (UTS) of 1114 MPa and 1055 MPa, respectively. The same can be found for the measured hardness of the steels which is also comparable at 332 HV30 and 343 HV30. A comparison of ductility and toughness of steels B1 and B2 shows that the uniform and total elongation are higher for steel B2 by 1.3% and 3.3%, while the impact energy is increased from 6 J to 15 J. The progression of the force–displacement curves of these materials, as displayed in Figure 12b, shows that both materials fracture without any plastic deformation. The impact energy calculated from the force–displacement curves ($CVN_c$) results in 4.4 J and 13.6 J for steels B1 and B2, respectively. B1 reaches a maximum force $F_{max}$ of 14.34 N, while B2 reaches 21.52 N.

The martensitic material shows higher strength values (YS: 847 MPa, UTS: 1341 MPa) and higher hardness in the air-cooled condition, as expected. The uniform elongation reaches 4%, which is slightly below the B1 and B2 material, while $A_t$ has 12.3% in between the values of B1 and B2. The impact energy is 37 J with an $F_m$ of 27.45 N. The annealing of the martensitic steel leads to an increase of the YS by 67 MPa and a decrease of the UTS by 60 MPa, while the elongations stay approximately the same ($A_u$: ±0.0%, $A_t$: +0.4%). An increase of the impact energy can also be observed after annealing (up to 44 J), while $F_{max}$ stays unchanged (27.55 N).

**Table 3.** Mechanical properties obtained by tensile tests, hardness measurements, and instrumented Charpy V-notch tests.

| Alloy | YS [mPa] | UTS [mPa] | $A_u$ [%] | $A_t$ [%] | $CVN_m$ [J] | $CVN_c$ [J] | $F_{max}$ [kN] | Hardness [HV30] |
|---|---|---|---|---|---|---|---|---|
| Q + T [7] | 959 | 1091 | 4.9 | 13.0 | — | — | — | — |
| B1 | 732 ± 22 | 1114 ± 28 | 4.6 | 11.6 | 6 | 4.4 | 14.35 | 332 |
| B2 | 721 ± 11 | 1055 ± 11 | 5.9 | 14.9 | 15 | 13.6 | 21.52 | 343 |
| M | 847 ± 23 | 1341 ± 3 | 4.0 | 12.3 | 37 | 30.5 | 27.45 | 406 |
| M + ba | 914 ± 25 | 1281 ± 1 | 4.0 | 12.7 | 44 | 36.3 | 27.55 | 372 |

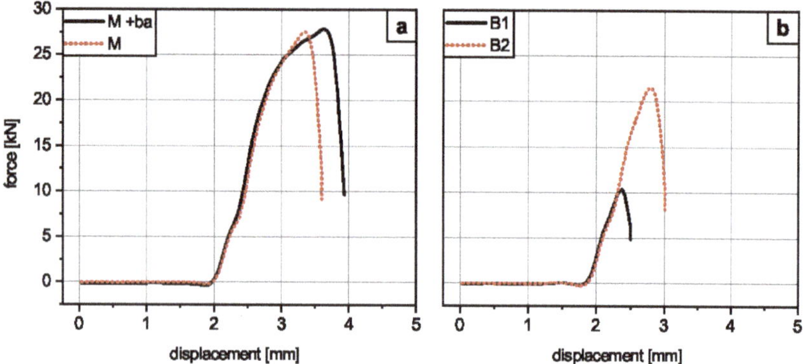

**Figure 12.** Force–displacement curves of the investigated materials obtained by instrumented Charpy V-notch impact test. The curves of the martensitic steel for both conditions are shown in (**a**), while the bainitic steels are displayed in (**b**).

## 4. Discussion

A comparison of the investigated steels shows that the bainitic grades differ concerning their ductility and toughness, while the strength is nearly identical. As the influence of prior austenite grains can be neglected (Figure 7), it can be concluded that the higher ductility and toughness of B2 can be attributed to the much finer bainite morphology. The difference in bainite morphology can be explained by the chemical composition of steels B1 and B2, especially by the carbon concentrations, which is much higher in alloy B1 than in alloy B2. Additionally, the steels contain different concentrations of boron and molybdenum, which are known to influence the bainite morphology [20]. Larger concentrations of carbon naturally enable higher phase fractions of carbon-containing bainitic microstructure constituents such as carbides and martensite–austenite (M–A) islands. Molybdenum is known to decrease the size of the (M–A) islands [20], and further depletes the matrix of carbon by precipitating MoC carbides [20,21]. The effect of boron is predominantly the suppression of diffusion-controlled phase transformations [22] and thus an increase in hardenability. In addition, boron decreases the proportion of coarse lath bainite [23], which can have positive effects on strength and toughness. The superstructures observed in Figure 3 are presumably caused by chemical inhomogeneities or the small differences in the degree of deformation caused by the hot-rolling process, as the features align with the rolling direction. The microstructure of the martensitic alloy in the as-forged condition is comparable to microstructures in different forging components which have been investigated in the past [9]. The microstructure consists of martensite and carbides, which demonstrates that in situ tempering takes place during air-cooling. The additional annealing at 450 °C for 2 h leads to an improvement of the yield ratio (from 0.63 to 0.71) and to an increase of the impact toughness. Both changes can be attributed to the coarsening of existing carbides, the resulting depletion of the matrix by carbon and the relaxation of the martensitic microstructure, as previously reported for alloys cast on the laboratory scale [24]. The progression of force–displacement curves of the bainitic grades (Figure 12) demonstrates that both materials fracture completely brittle. This corresponds well with the observed fracture surfaces (Figure 8) which only show brittle fracture. The differences in impact energy (+9 J) can be explained by the higher $F_{max}$, which was reached during the tests. The force–displacement curves of both sample states show ductile deformation (Figure 12a), with steel M + ba having slightly more deformation before the fracture. This also corresponds well with the observed fracture surfaces (Figure 8), which show a mixture of brittle and ductile regions.

Statistical analyses revealed that the prior austenite grain size of the martensitic alloys has a broader scattering than the bainitic alloys. This is demonstrated by the size of the interquartile ranges as well as the width of Weibull fit of the histograms. TEM analyses of chemically similar alloys showed that hot forging at 1200 °C with consecutive air cooling results in niobium carbide precipitates with sizes between 20 nm to 50 nm [8]. Niobium precipitates are known for the retardation of recrystallization [25]. However, thermodynamic calculations with MatCalc6 reveal that NbC precipitates in steel M dissolve at 1100 °C. Therefore, it can be assumed that the niobium precipitates are dissolved after the hot forging (at least to a certain extend), which might reduce their beneficial effect.

The EDX measurements reveal the presence of manganese sulfides (MnS) in the bainitic materials as well as in the martensitic materials. Aluminum nitrides (AlN) were naturally only observed in the martensitic steel, as the aluminum concentration is much higher in comparison to the bainitic grades. The additional increase of aluminum and nitrogen in scan 3 can be explained by a second AlN precipitated close to or beyond the primary investigated MnS-particle. The negative effect of MnS particles on the impact toughness caused by them acting as inner notches is well known in the literature [26]. Especially for steel products which exhibited large degrees of deformation during the production, elongated MnS particles might lead to anisotropic impact toughness [27–29]. As reported, large MnS particles were observed in the microstructure (Figure 11) and at the fracture surfaces (Figure 10) of the martensitic steel, accompanied by AlN precipitations in the close vicinity. Similar observations have been reported by Marich and Player [30], who found that $Al_2O_3$ inclusions act as nucleation sites for MnS, which was also observed for MgO in magnesium killed steels [31]. It can therefore be assumed that AlN acts similar in the investigated material, explaining the relatively large MnS-particles in the martensitic steel. The bainitic steels exceed the strength of standard PHFP steels in the air-cooled condition [5], while the impact toughness is similar. The higher strength of the bainitic steels might enable new lightweight designs of con-rods, which are currently produced from PHFP steels. As the Charpy tests revealed completely brittle failure, for the investigated bainitic steels, it can be assumed that the splitting process can be performed successfully. The martensitic steel achieved higher strengths (YS: 847 MPa) and the impact toughness reached values above 30 J, which is required for most automotive chassis application [9]. However, due to the partly ductile fracture, quality issues might occur during the splitting process [17,18]. Both materials offer the possibility to reduce the $CO_2$-footprint of con-rods, either by lightweighting or by a reduction of the heat treatment, if quench and tempering steels are substituted. Lightweighting of the con-rods will additionally reduce the necessary counterweights at the crankshaft as well as other motor components [11].

## 5. Conclusions

Air-hardening steels offer mechanical properties close to those of standard quench and tempering steels. A broad market introduction of these steels can reduce the carbon footprint of the forging industry, as the heat treatment process is significantly shortened. Especially for die-forged con-rods, these materials show interesting combinations of strength and toughness, as the production of these components requires rather low toughness values. The following can be concluded from this study:

- The presented steels show good balances of strength and ductility in the air-hardened condition, which enables the substitution of other standard steel grades.

- The $CO_2$-footprint of forged components can be reduced by lightweighting or shortening the heat treatment through the utilization of new air-hardening materials.
- The combination of impact toughness and yield strength is suitable for con-rod production in the case of the bainitic steels, while final confirmation has to be made for the martensitic steel.
- A comparison of the bainitic grades reveals that the 19MnCrMo7-6 steel achieves comparable strength but higher elongations and impact toughness than the 35MnCrB6-4 steel. The differences in mechanical properties can be explained by the differences in the bainite morphology.

**Author Contributions:** Conceptualization, A.G.; methodology, A.G.; formal analysis, A.G. and K.B.; investigation, A.G. and K.B.; resources, A.G., R.L., and U.Z.; data curation, A.G.; writing—original draft preparation, A.G. and K.B.; writing—review and editing, R.L. and U.Z.; visualization, A.G. and K.B. All authors have read and agreed to the published version of the manuscript.

**Funding:** This research received no external funding.

**Institutional Review Board Statement:** Not applicable.

**Informed Consent Statement:** Not applicable.

**Data Availability Statement:** The data presented in this study are available on request from the corresponding author.

**Conflicts of Interest:** The authors declare no conflicts of interest.

# References


1. Bleck, W.; Bambach, M.; Wirths, V.; Stieben, A. Microalloyed Engineering Steels with Improved Performance—An Overview. *HTM J. Heat Treat. Mater.* **2017**, *72*, 346–354. [CrossRef]
2. Wirths, V.; Wagener, R.; Bleck, W.; Melz, T. Bainitic Forging Steels for Cyclic Loading. *Adv. Mater. Res.* **2014**, *922*, 813–818. [CrossRef]
3. Sugimoto, K.I.; Sato, S.H.; Kobayashi, J.; Srivastava, A.K. Effects of Cr and Mo on Mechanical Properties of Hot-Forged Medium Carbon TRIP-Aided Bainitic Ferrite Steels. *Metals* **2019**, *9*, 1066. [CrossRef]
4. Hojo, T.; Kobayashi, J.; Sugimoto, K.i.; Nagasaka, A.; Akiyama, E. Effects of Alloying Elements Addition on Delayed Fracture Properties of Ultra High-Strength TRIP-Aided Martensitic Steels. *Metals* **2020**, *10*, 6. [CrossRef]
5. Sugimoto, K.i.; Hojo, T.; Srivastava, A.K. Low and Medium Carbon Advanced High-Strength Forging Steels for Automotive Applications. *Metals* **2019**, *9*, 1263. [CrossRef]
6. Keul, C.; Wirths, V.; Bleck, W. New bainitic steels for forgings. *Arch. Civ. Mech. Eng.* **2012**, *12*, 119–125. [CrossRef]
7. Stieben, A.; Bleck, W.; Schönborn, S. Lufthärtender duktiler Stahl mit mittlerem Mangangehalt für die Massivumformung. *massivUmformung* **2016**, *1*, 50–55.
8. Gramlich, A.; Schmiedl, T.; Schönborn, S.; Melz, T.; Bleck, W. Development of air-hardening martensitic forging steels. *Mater. Sci. Eng. A* **2020**, *784*, 139321. [CrossRef]
9. Gramlich, A.; Schönborn, S.; Schmiedl, T.; Baumgartner, J.; Krupp, U. Lufthärtende duktile Schmiedestähle für zyklische Beanspruchung. *massivUmformung* **2021**, *5*, 64–69.
10. Wegner, K.W. Werkstoffentwicklung für Schmiedeteile im Automobilbau. *ATZ—Automob. Z.* **1998**, *100*, 918–927. [CrossRef]
11. Lapp, M.T.; Hall, C.C. Verringerung bewegter Massen durch Leichtbaupleuel. *Lightweight Des.* **2011**, *4*, 30–37. [CrossRef]
12. Spangenberg, S.; Kemnitz, P.; Kopf, E.; Repgen, B. Massereduzierung an Bauteilen des Kurbeltriebs. *MTZ—Mot. Z.* **2006**, *67*, 254–261. [CrossRef]
13. Lipp, K.; Kaufmann, H. Schmiede- und Sinterschmiedewerkstoffe für Pkw-Pleuel. *MTZ—Mot. Z.* **2011**, *72*, 416–421. [CrossRef]
14. Shi, Z.; Kou, S. Inverse Reconstruction of Fracture Splitting Connecting Rod and its Strength and Fatigue Life. *J. Fail. Anal. Prev.* **2018**, *18*, 619–627. [CrossRef]
15. Fukuda, S. Development of fracture splitting connecting rod. *JSAE Rev.* **2002**, *23*, 101–104. [CrossRef]
16. Gu, Z.; Yang, S.; Ku, S.; Zhao, Y.; Dai, X. Fracture splitting technology of automobile engine connecting rod. *Int. J. Adv. Manuf. Technol.* **2005**, *25*, 883–887. [CrossRef]
17. Shi, Z.; Kou, S. Study on Fracture-Split Performance of 36MnVS4 and Analysis of Fracture-Split Easily-Induced Defects. *Metals* **2018**, *8*, 696. [CrossRef]
18. Shi, Z.; Kou, S. Analysis of quality defects in the fracture surface of fracture splitting connecting rod based on three-dimensional crack growth. *Results Phys.* **2018**, *10*, 1022–1029. [CrossRef]
19. Smith, R.L.; Sandly, G.E. An Accurate Method of Determining the Hardness of Metals, with Particular Reference to Those of a High Degree of Hardness. *Proc. Inst. Mech. Eng.* **1922**, *102*, 623–641. [CrossRef]

20. Ackermann, M.; Resiak, B.; Buessler, P.; Michaut, B.; Bleck, W. Effect of Molybdenum and Cooling Regime on Microstructural Heterogeneity in Bainitic Steel Wires. *Steel Res. Int.* **2020**, *37*, 1900663. [CrossRef]
21. Capdevila, C.; Caballero, F.G.; García De Andrés, C. Determination of Ms Temperature in Steels: A Bayesian Neural Network Model. *ISIJ Int.* **2002**, *42*, 894–902. [CrossRef]
22. Gramlich, A.; van der Linde, C.; Ackermann, M.; Bleck, W. Effect of Molybdenum, Aluminium and Boron on the phase transformation in 4 wt.–% Manganese Steels. *Results Mater.* **2020**, *8*, 100147. [CrossRef]
23. Wang, X.M.; He, X.L. Effect of Boron Addition on Structure and Properties of Low Carbon Bainitic Steels. *ISIJ Int.* **2002**, *42*, S38–S46. [CrossRef]
24. Gramlich, A.; Bleck, W. Tempering and Intercritical Annealing of Air-Hardening 4 wt% Medium Manganese Steels. *Steel Res. Int.* **2021**, *92*, 2100180. [CrossRef]
25. Speer, J.G.; Hansen, S.S. Austenite recrystallization and carbonitride precipitation in niobium microalloyed steels. *Metall. Mater. Trans. A* **1989**, *20*, 25–38. [CrossRef]
26. Cyril, N.; Fatemi, A.; Cryderman, B. Effects of Sulfur Level and Anisotropy of Sulfide Inclusions on Tensile, Impact, and Fatigue Properties of SAE 4140 Steel. *SAE Int. J. Mater. Manuf.* **2009**, *1*, 218–227. [CrossRef]
27. Pickering, F.B. Some Effects of Mechanical Working on the Deformation of Non-Metallic Inclusions. *J. Iron Steel Inst.* **1958**, *189*, 148–159.
28. Berns, H. Zur Zähigkeit von Vergütungsstählen. *Mater. Werkst.* **1978**, *9*, 189–204. [CrossRef]
29. Kirby, B.G.; van Tyne, C.J.; Matlock, D.K.; Krauss, G.; Turonek, R.; Filar, R.J. *Carbon and Sulfur Effects on Performance of Microalloyed Spindle Forgings*; SAE Technical Paper Series; SAE International: Warrendale, PA, USA, 1993. [CrossRef]
30. Marich, S.; Player, R. Sulfide inclusions in iron. *Metall. Mater. Trans. B* **1970**, *1*, 1853–1857. [CrossRef]
31. Kimura, S.; Nakajima, K.; Mizoguchi, S.; Hasegawa, H. In-situ observation of the precipitation of manganese sulfide in low-carbon magnesium-killed steel. *Metall. Mater. Trans. A* **2002**, *33*, 427–436. [CrossRef]

*Article*

# Redistribution of Grain Boundary Misorientation and Residual Stresses of Thermomechanically Simulated Welding in an Intercritically Reheated Coarse Grained Heat Affected Zone

Giancarlo Sanchez Chavez [1,*], Segen Farid Estefen [2], Tetyana Gurova [3], Anatoli Leontiev [4], Lincoln Silva Gomes [5] and Suzana Bottega Peripolli [5]

[1] Ingeniería de Materiales, Universidad Nacional de San Agustín de Arequipa, Arequipa 04001, Peru
[2] Laboratório de Tecnologia Submarina, Instituto Alberto Luiz Coimbra de Pós-Graduação e Pesquisa de Engenharia, Universidade Federal do Rio de Janeiro, Rio de Janeiro 21941-901, RJ, Brazil; segen@lts.coppe.ufrj.br
[3] Curso de Tecnologia em Construção Naval, Fundação Centro Universitário Estadual da Zona Oeste, Rio de Janeiro 23070-200, RJ, Brazil; gurova@lts.coppe.ufrj.br
[4] Instituto de Matemática, Universidade Federal do Rio de Janeiro, Rio de Janeiro 21941-901, RJ, Brazil; anatoli@im.ufrj.br
[5] Inspeção e Integridade, Instituto SENAI de Inovação, Rio de Janeiro 20550-011, RJ, Brazil; lsgomes@firjan.com.br (L.S.G.); speripolli@firjan.com.br (S.B.P.)
* Correspondence: gsanchezch@unsa.edu.pe; Tel.: +51-957822222

**Citation:** Sanchez Chavez, G.; Farid Estefen, S.; Gurova, T.; Leontiev, A.; Silva Gomes, L.; Bottega Peripolli, S. Redistribution of Grain Boundary Misorientation and Residual Stresses of Thermomechanically Simulated Welding in an Intercritically Reheated Coarse Grained Heat Affected Zone. *Metals* **2021**, *11*, 1850. https://doi.org/10.3390/met11111850

Academic Editor: Koh-ichi Sugimoto

Received: 13 October 2021
Accepted: 15 November 2021
Published: 18 November 2021

**Publisher's Note:** MDPI stays neutral with regard to jurisdictional claims in published maps and institutional affiliations.

**Abstract:** A study of the migration of the grain boundary misorientation and its relationship with the residual stresses through time immediately after the completion of a thermomechanical simulation has been carried out. After physically simulating an intercritically overheated welding heat affected zone, the variation of the misorientation of grain contours was observed with the electron backscatter diffraction (EBSD) technique and likewise the variation of the residual stresses of welding with RAYSTRESS equipment. It was observed that the misorientation of the grain contours in an ASTM DH36 steel was modified after the thermomechanical simulation, which corresponds to the measured residual stress variation along the first week of monitoring, with compressive residual stresses ranging from 195 MPa to 160 MPa. The changes in misorientation indicate that the stress relaxation phenomenon is associated with the evolution of the misorientation in the microstructure caused by the welding procedure. On the first day, there was a fraction of 4% of the kernel average misorientation (KAM) values at 1° misorientation and on the fourth day, there was a fraction of 7% of the KAM values at 1° misorientation.

**Keywords:** grain boundary misorientation; welding residual stresses; dislocations

## 1. Introduction

Fusion welding, despite being one of the most used technologies in the merging of metals such as high-strength steels, has a considerable number of drawbacks, among which the internal stresses caused by the transient thermal cycles suffered in welds of various steps—which form part of the procedures for large structures, such as vessels for oil extraction—stand out. Thermal stresses, residual stresses, and distortions can cause fractures in the areas around welds under certain tensile load conditions [1]. In addition to these, stress corrosion and fatigue problems are also related to these residual stresses. Residual stresses and corrosion, which affect the lifetime of structural components working at high temperatures, must be considered [2]. The compressive residual stresses in a 316L stainless steel increase in the resistance to pit initiation [3]. The cavitation erosion pits have higher stress levels which lead to fatigue crack nucleation [4]. When a material has low levels of yield strength, the residual stresses complicate the performance of the welded joint, reducing its strength. Due to the particularity of welding processes, the mechanical properties of welded joint materials, especially the yield strength, are unevenly distributed, which seriously affects the safety performance of welded joints [5]. The heat softening zone has a great reduction in strength compared with the base metal when lap joints of 1180 MPa steel sheets are welded [6]. For example, the magnitude of tensile residual stresses reaches the yield strength of base metal; this is because, during the welding, the filler wire and base metal experience rapid heating and a quick cooling process [7].

By monitoring the residual stresses of weld daily by means of X-ray diffraction, magnetic methods, and vertical displacements of the surface of the welded plate using laser technology equipment, the redistribution of the residual stresses was observed in a relatively short period of up to two weeks following the welding procedure [8–10]. This observable stress redistribution, not observed in other research works, is characterized by a reduction in and uniformity of the maximum shear stress values 2 weeks after welding. Microstructural analysis rules out the possibility of stress redistribution owing to the failure of the material.

Changes in residual stresses over time are related to the misorientation of grain boundaries. Internal stresses are related to the microstructure on an atomic scale and are present in areas close to dislocations [1]. Researchers such as Taylor [11], Orowan [12], and Love [13] have related elastic and plastic distortions and/or deformations to the dislocations present in the microstructure of materials. The Volterra process shows that the creation of a dislocation requires energy and the introduction of residual stresses [14]. Fan et al. [15] state that the dislocation glide is a general mode of strain, governing the strength of metals, so there is a dependence on the rate of strain and the density of a dislocation with the dynamics of collective dislocation. When there is plastic strain, the distortion of the crystal lattice is alleviated with the formation of dislocations with a certain concentration of dislocations, with a net Burger vector other than zero, that cause changes in the orientation of the crystal lattice [16]. Relating residual stresses at the atomic and microscopic level, during plastic strain individual grains do not twist as a single element in magnitude and direction; thus, dislocations are located in the grain contours to maintain the continuity of the microstructural network [17]. For metals deformed to moderate strains, the dislocations density increases linearly with plastic strain [18]. In steels, the deformation introduces dislocation boundaries with a small misorientation in austenite [19,20]. The behavior of the dislocations within the microstructure is heterogeneous [21]. Hence, the dislocations are highly heterogeneously distributed, with significant accumulation of high density near the grain boundaries and relatively low density within the interior of the grain [22]. The orientations of the crystals within the grains change due to the accumulation of dislocations and may fluctuate by several degrees, even within the same grain [23]. Therefore, the accumulation of dislocations is expected to change the grain orientation distribution and the grain boundary misorientation angle. This is also based on the fact that the nature of any given grain contour depends on the misorientation of the two adjacent grains and the orientation of the boundary plane with respect to them [24].

The level of deformation can be estimated with misorientation analysis, where the orientation of two or more data points is compared for individual grains or within calculational domains [25]. The nature of any given grain contour depends on the misorientation of the two adjacent grains and the orientation of the boundary plane in relation to them [24]. Misorientation within grains increases with the introduction of plastic strain, which is often used as a quantitative measure to describe the degree of microstructure deformation [26]. However, the introduction of plastic strain can result in substructure formation associated with the rearrangement of dislocations [26]. Because of this, small misorientations are important for understanding the mechanical behavior of materials and it is important that the orientation measurements are precise [27,28]). The yield stress, as well as the flow stress after yielding, increases as the grain boundary misorientation angle increases [29]. When a deformation is applied to a metal, the presence of low-angle grain boundaries (LAGB) is observed [19]. Free dislocations generated during plastic deformation can readily rearrange themselves, leading to the development of low-angle grain boundaries [26]. Further deformation is enabled by the formation of new high-angle grain boundaries [25]. In a work published by Costa et al. [30], for low-angle grain boundaries, samples with higher stress levels have a higher percentage of misorientation, and for high-angle grain boundaries the observed behavior is the opposite. This is due to the fact that the energy of the low-angle grain boundaries is the total energy of the dislocations within the contour area and depends on the spacing of the dislocations and the misorientation angle, whereas, for high-angle grain boundaries, the grain boundary energy is almost independent of the misorientation [24]. Humphreys et al. [31] state that low-angle boundaries are those that are composed of an arrangement of dislocations whose structure and properties vary depending on the misorientation, while high-angle boundaries are those whose structure and properties generally do not depend on misorientation. This suggests that there is a direct relationship between the stress–strain level and the amount of grain boundaries, with the low-angle grain boundaries having a greater influence on the stress–misorientation relationship. From what was discussed in previous paragraphs, a hypothesis was proposed: the stress relaxation phenomenon is associated with the evolution of the misorientation of the microstructure that occurs in the material following the culmination of the welding procedure.

## 2. Materials and Methods

A 12 mm thick ASTM DH36 steel sheet was used, from which specimens with dimensions of 10 mm × 10 mm × 71 mm were extracted (Figure 1). In the center of this sample was the intercritically reheated coarse grained heat affected zone (IC CGHAZ), and next to it was the fine grain heat affected zone (FGHAZ). The elastic limit of the material is 360 MPa. The chemical composition of the steel was obtained in this investigation with optical emission spectroscopy using a Foundry Master Pro spectrometer from Oxford Instruments (Concord, MA, USA), and is given in Table 1 with a carbon equivalent ($CE_{iiw}$) of 0.41%, indicating that the solubility is relatively good since it exceeds 0.40% of $CE_{iiw}$. Steels with $CE_{iiw}$s < 0.40% are considered to have good weldability and do not require special preparation or post-treatment of the weld [32]. The equation for the calculation of the $CE_{iiw}$ was formulated by the International Institute of Welding (IIW) [32].

**Figure 1.** Simulated specimen with dimensions of 10 mm × 10 mm × 71 mm presenting the IC CGHAZ zone that is located in the central part of the sample with a width of approximately 8 mm.

**Table 1.** Chemical composition of the ASTM DH36 steel in wt.%.

| C% | Mn% | Si% | Cr% | Cu% | Ni% | Mo and Ti% | V% | Nb% | P% | S% | $CE_{iiw}$% |
|---|---|---|---|---|---|---|---|---|---|---|---|
| 0.141 | 1.49 | 0.179 | 0.022 | 0.0106 | 0.0071 | <0.0005 | 0.0429 | 0.0322 | 0.0224 | 0.0102 | 0.41 |

The thermal welding cycles were carried out with a physical simulation for the determination of the microstructure using Gleeble® 3800 equipment (Dynamic Systems Inc., Poestenkill, NY, USA). The sample was held rigidly in the jaws of the machine, not allowing the jaws to move when the material presented volumetric expansion, as shown in Figure 2. The temperature was measured using a type K thermocouple.

**Figure 2.** Specimen (in the center of the image) attached to the Gleeble® 3800 equipment.

The thermal cycle simulation of double pass welding consisted of the following steps. For the first welding pass simulation, the middle region of the test specimen was heated to 1350 °C with a heat rate of 450 °C/s, was maintained at this temperature for 0.35 s, was cooled down to 800 °C in 20 s with a cooling rate of 27.5 °C/s, and then to 200 °C in $t_{8/5}$ = 160 s with a cooling rate of 3.75 °C/s. The second pass simulation involved heating from 200 °C to a peak temperature of 800 °C with a heat rate of 266 °C/s, maintaining the specimen at this temperature for 0.35 s, and then cooling it to 200 °C in $t_{8/5}$ = 160 s with a cooling rate of 3.75 °C/s followed by air free cooling, as shown in Figure 3.

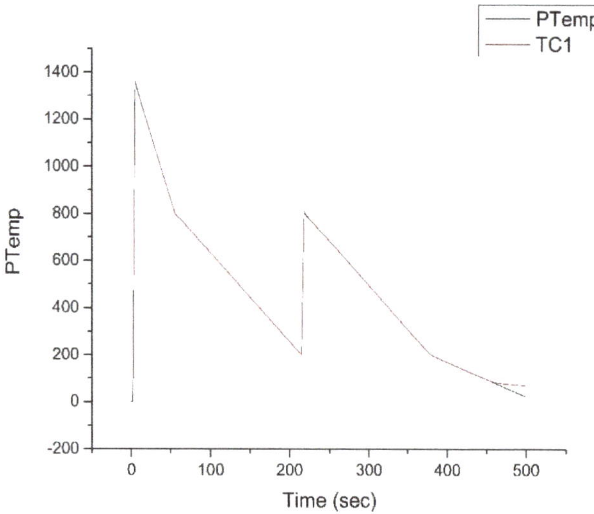


**Figure 3.** Thermal cycles simulating two welding passes to generate the IC CGHAZ zone.

These thermal cycles were designed to obtain the microstructure of a heat affected zone (HAZ) from two welding passes. The first pass reaches a coarse-grained heat affected zone and the second pass pertains to an intercritical zone (IC CGHAZ). The temperature of the second intercritical pass, which was 800 °C, was taken between the temperatures AC1 = 695 °C and AC3 = 830 °C calculated with the Thermo-Calc® (Thermo-Calc Software's, Pittsburgh, USA) version 1 thermodynamic software based on TCS Steel and Fe-alloys Database (TCFE) data (Figure 4).

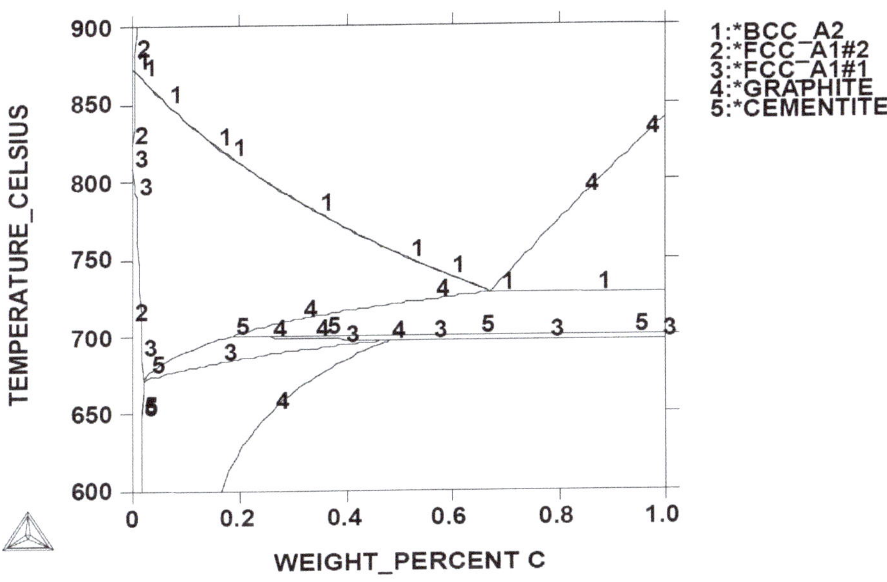


**Figure 4.** Intercritical zone of DH36 steel calculated with Thermo-Calc®.

For data acquisition, an FEI Quanta 450 scanning electron microscope (SEM, FEI Technologies Inc., Hillsboro, Oregon, USA) with a tungsten filament and a BRUKER E-FLASH EBSD automatic detector (Bruker Nano GmbH, Berlin, Germany) were used. The BRUKER ESPRIT 2.0 package was applied for data processing. The configuration used for the microscope had an acceleration voltage of 23 KV, with a step size of 0.72 μm and an indexing speed (min–max.) of 94.6–97.9%. The samples were prepared using SiC paper with a grain from 100 to 1000 for grinding. Polishing was carried out with a diamond suspension from 6 μm to 1 μm, and finally a fine polishing was applied using a silica colloidal suspension of 0.25 μm.

The residual stress values were measured using a RAYSTRESS instrument (SYN-THESIS Co. ltd, Saint-Petersburg, Russia), which is a portable X-ray machine that employs a double exposure method [8,9]. The principle of the stress measurements via the RAYSTRESS using double exposure is shown in Figure 5. Two cassette windows captured the diffraction lines in 2θ-angular intervals from 148° to 164°. The inclination of the specimen surface of 12° corresponds to measurements for steel specimens using Cr-Kα radiation and the {211} reflection with θ211 = 78°. The experimental accuracy of the stress measurements was 10 MPa. The specimen was maintained at the same measurement position throughout the entire three-week monitoring period. Stress measurements were taken from the center of the specimen in the IC CGHAZ area, and in the area located 5 mm from the center of the specimen, whose microstructure was identified as FGHAZ, which is the fine grain heat affected zone. The measurements on both areas were performed simultaneously using two RAYSTRESS machines. Measurements were started around two hours after the end of the Gleeble simulation and were performed every 12 h.

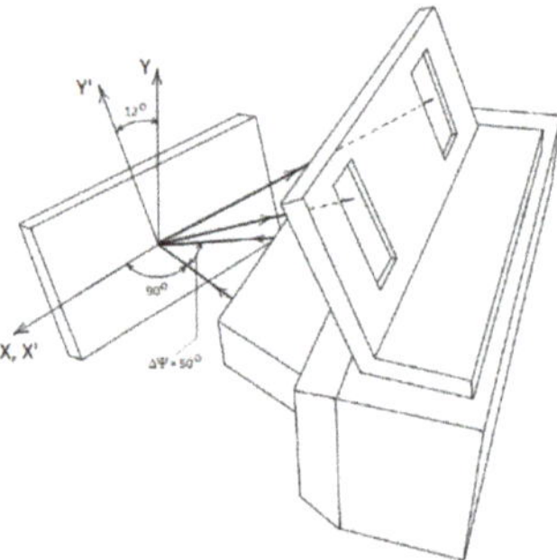

**Figure 5.** Schematics of the stress measurements using the RAYSTRESS machine.

## 3. Results

### 3.1. Microstructure of Base Material Obtained by Thermomechanical Simulation

The microstructure of the base material without thermomechanical simulation obtained with the EBSD technique is shown in Figure 6. Figure 6a shows the inverse pole images (IPF) indicating the crystallographic directions caused by the ferrite, austenite, upper bainite, and cementite. A predominant direction was not evident since orientations of the type 001, 101, 111, and 210 were present throughout the microstructure. In Figure 6b, a predominantly ferritic structure can be observed. The dark regions represent the banding resulting from the thermomechanical treatment of the sheet (Figure 6c), which was composed of phases belonging to bainite, austenite, and cementite as shown in Figure 6b. These phases can also be observed in certain sections of the grain boundaries, but in lesser amounts. The colored areas of the EBSD micrographs in Figure 6a–c that contain the cementite, bainite, and austenite phases are related to the pearlitic and grain boundary zone of Figure 6d. So, it could be said that in the banded areas of the base material, we can find retained austenite that could be forming the martensite–austenite constituent (MA). The microconstituent MA with different morphologies is located mainly in the contours of the anterior austenitic grain, the lathes of other phases, and rarely within the grains [33]. According to the work of Moeinifar et al. [34], the mean diameter of the MA microconstituents in their largest size is, on average, 0.93 µm. MA blocks can be approximately 3 to 5 µm in size according to Davis and King [35] or 1 µm according to You et al. [36].

### 3.2. Microstructure IC CGHAZ Simulated Thermomechanically

The microstructure of the central part of the thermomechanically simulated sample (Figure 1) is a microstructure that would belong to a multipass region of real welding, called IC CGHAZ, which is observed in Figure 7. In the inverse pole figure (Figure 7a), a predominance of crystallographic orientations of 101 for ferrite and austenite, of 110 for upper bainite, and 010 for cementite is observed. In Figure 7a,b, within the initial austenitic grains, phases in elongated lathes are observed oriented according to the crystallographic directions 110 and 210 belonging to a bainitic structure. The phase map of Figure 7c indicates that 95.2% belongs to the ferritic phase. From Figure 7d, we can see that the microstructure belongs to a bainitic structure in the form of blocks with dark phases between the strips of the bainite, phases that belong to austenite, cementite, and bainite as indicated by the phase map in Figure 7c. Steels with low carbon content contain cementite through the matrix as a second dispersed phase [37], due to the excess solubility of carbon in the ferrite.

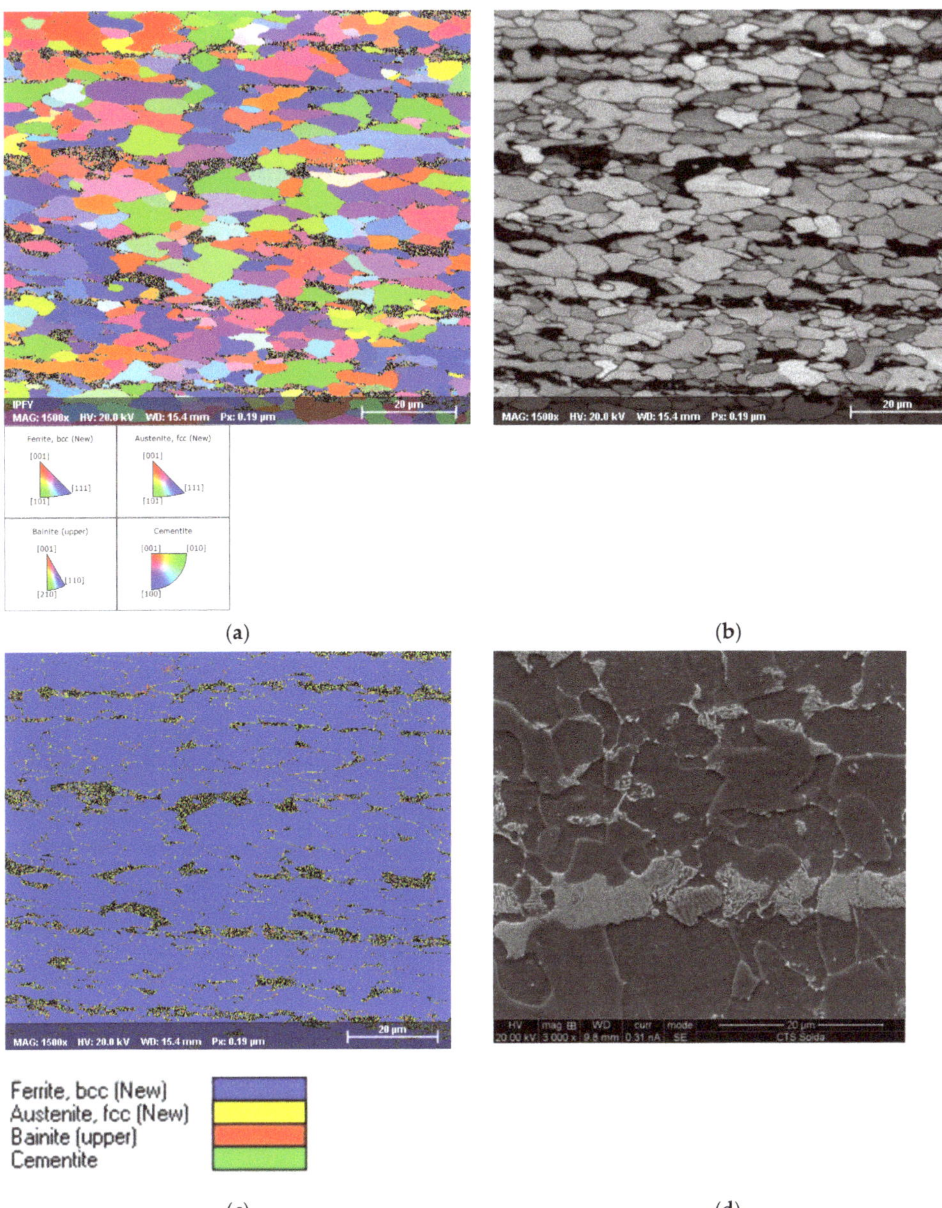

**Figure 6.** Micrographs of the base material obtained with EBSD: (**a**) inverse pole figure (IPF) normal to the Y axis, (**b**) map of the quality standard, and (**c**) phase distribution map. Magnitude of 1500×. (**d**) Electron microscopy micrograph with magnitude of 3000×.

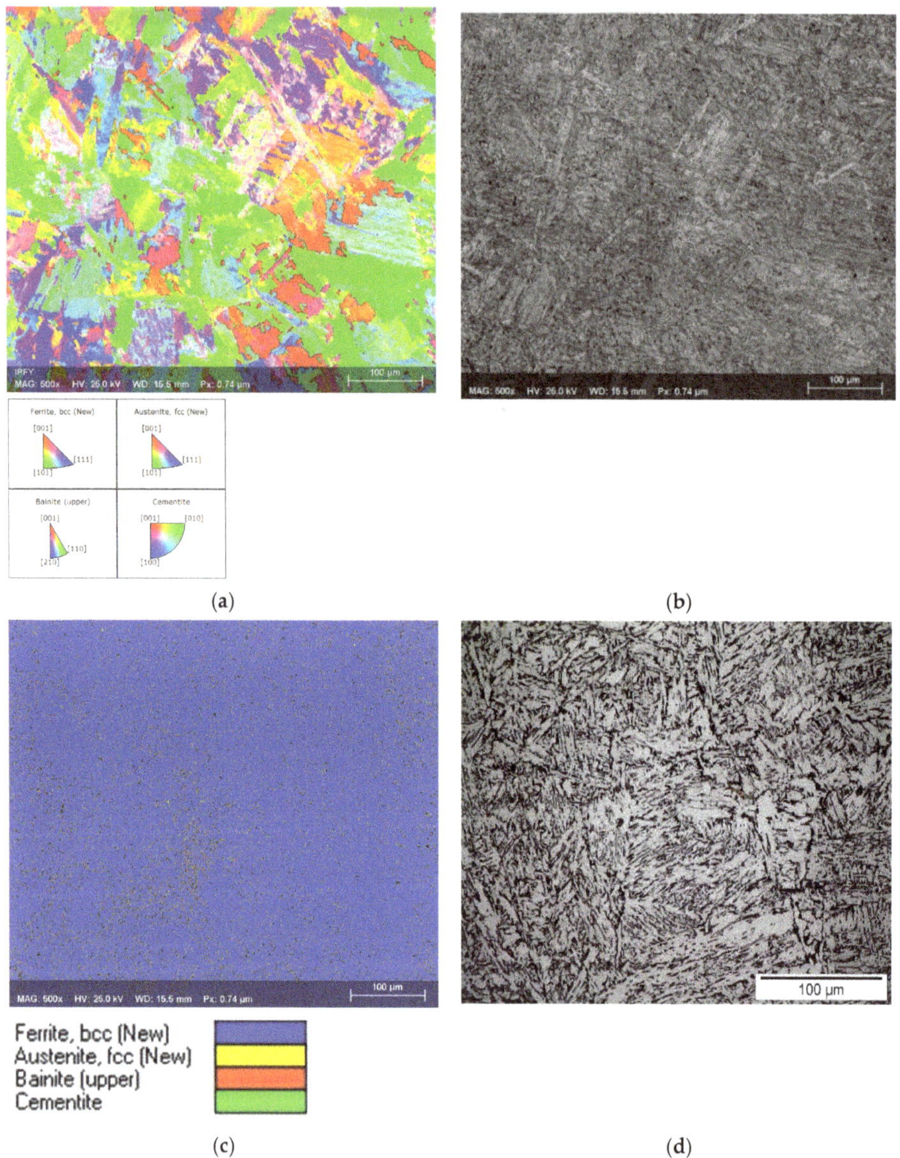

**Figure 7.** IC CGHAZ micrographs obtained with EBSD: (**a**) inverse pole figures (IPFs) normal to the Y axis, (**b**) quality pattern map (PQ), and (**c**) phase distribution maps. (**d**) Light microscopy micrograph. Magnitude of 500×.

In Figure 7c, black dots are seen representing 1.37% of zero solutions, which are non-indexed patterns that can indicate an MA microconstituent, precipitates, carbides, or second phases that are located on the edges of the initial austenitic grain and dark phases of a smaller size within them. It was not possible to obtain a good quality of Kikuchi standards from the carbides due to their discontinuities and a high density of dislocations at the respective interfaces [38]. These black areas are regions of higher dislocation density and high residual stresses introduced by significant differences in the coefficient of thermal expansion between the reinforcing particles and the matrix, resulting in a strong dislocation field around the precipitates [39]. These dislocations are regions of greater misorientation that are difficult to index and therefore appear as non-indexed black dots [40].

*3.3. Redistribution of the Residual Tension*

Figure 8 and Table 2 shows the results of residual stress measurements performed on IC CGHAZ and FGHAZ, adjacent to the IC CGHAZ zone areas of the specimen over the 14 monitoring days. Sign "-" denotes compressive stress. The measurements were performed in the longitudinal direction. Although the stress state on the whole is defined as a tensor value, these measurements are sufficient to detect the change of the residual stress state on the analyzed areas for the purposes of this paper.

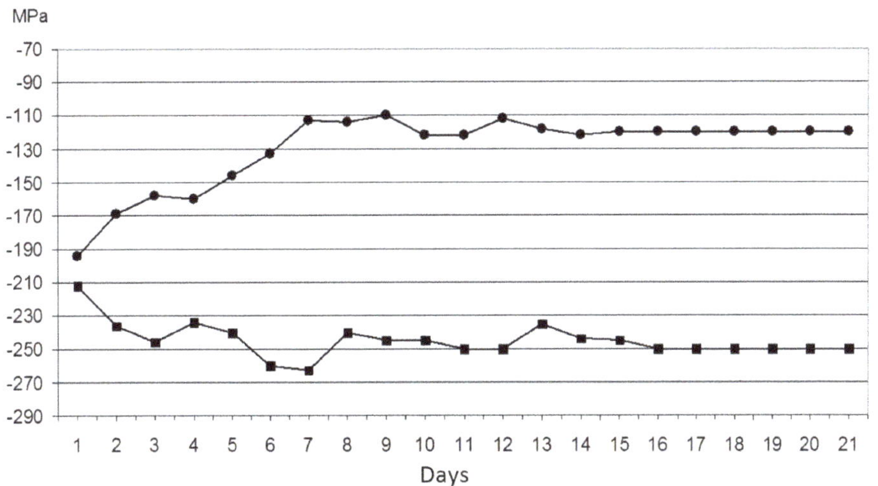

**Figure 8.** Change in residual stress over time. Plot with ● corresponds to the residual stresses measured on IC CGHAZ area; plot with ■ corresponds to the residual stresses measured on FGHAZ area.

**Table 2.** Values of the change in residual stress over time.

| Days | 1 | 2 | 3 | 4 | 5 | 6 | 7 | 8 | 9 | 10 | 11 | 12 | 13 | 14 |
|---|---|---|---|---|---|---|---|---|---|---|---|---|---|---|
| IC CG-HAZ (MPa) | −194 | −169 | −158 | −160 | −147 | −132 | −112 | −114 | −110 | −123 | −123 | −111 | −118 | −122 |
| FGHAZ (MPa) | −212 | −238 | −247 | −234 | −242 | −262 | −264 | −242 | 246 | 246 | −250 | −250 | −235 | −244 |

We observed a decrease in compressive residual stress in the IC CGHAZ area accompanied by an increase of compressive residual stress in the FGHAZ area. This process is not monotonic and has some variations. The initial difference in the values of the stresses on two analyzed areas was 20 MPa, and the final difference was 130 MPa. Stabilization in the process of stress redistribution usually begins from the second week. A two-week stabilization period is normally applied in real welding samples [8–10].

In Figure 8, we can see that, in the first days of monitoring, the compressive residual stresses measured in the longitudinal direction in the IC CGHAZ area had a value of 195 MPa and four days later this value became 160 MPa, which represents the absolute value of the residual stress decrease from the first day to the fourth day of observation. In the following sections, we will describe how the fractions of the misorientation values vary from the first to the fourth day after having completed the simulation of the heat affected zone, just as the residual stresses calculated with the X-ray diffraction method vary.

### *3.4. Redistribution of Misorientation in Microstructure*
### 3.4.1. Redistribution of KAM

The kernel average misorientation (KAM) maps represent the mean misorientation ratio of a given point and the mean orientation of its neighbors with the same grain [41,42]. KAM is designed to show very close local orientation changes. Analysis of the results based on the KAM parameters are presented in Figures 9 and 10. The KAM results give us up to an approximate value of 5 degrees of misorientation and these are considered as low-angle grain contours (LAGBs). This limit on misorientation values was chosen to elucidate the misorientation between the grain boundaries and interior grains [26]. A visible variation of the maximum values of the fraction of the misorientation angles was observed from the first to the fourth day of observation. For example, in Figure 10 it can be seen that, on the first day, there was a fraction of 4% of the KAM values at 1° misorientation and on the fourth day there was a fraction of 7% of the KAM values at 1° misorientation. In heat-affected areas of welded joints, when temperature gradients occur, different internal strains are noticeable, such strains being related to microstructural changes. The presence of LAGB is also considered to be due to incomplete recrystallization [43]. In our case, the redistribution of the residual strain of the IC CGHAZ zone was observed. These areas that are close to the weld beads are subject to higher cooling speeds and consequently large internal strains are present, as seen in Figure 9a, which, after the passage of time, gradually decreased, as can be seen in Figure 9b.

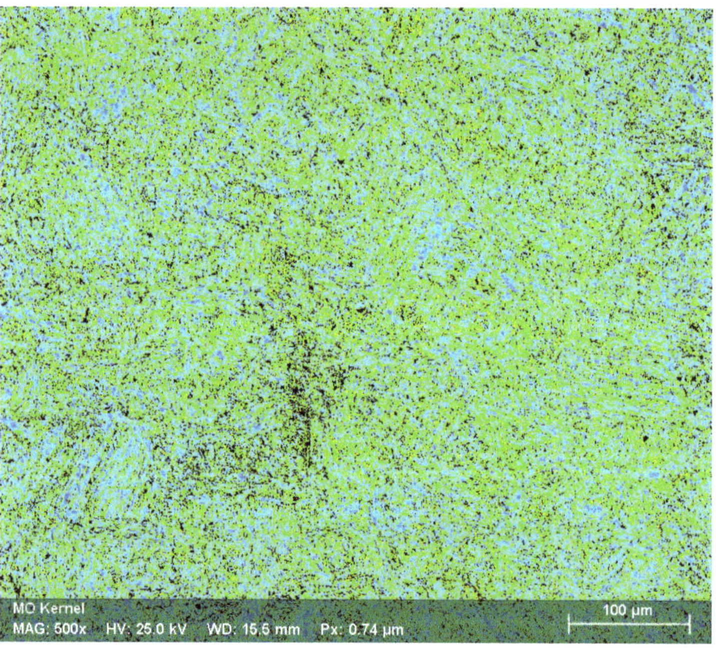


(a)

**Figure 9.** *Cont.*

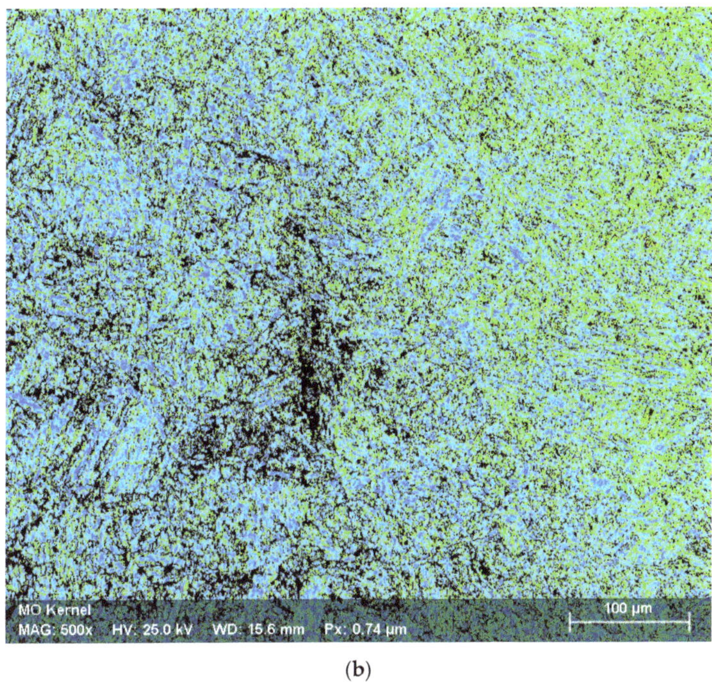


(**b**)

**Figure 9.** Maps of kernel average misorientation (KAM): (**a**) first day of monitoring KAM map and (**b**) fourth day of monitoring KAM map. EBSD with magnitude of 500×.

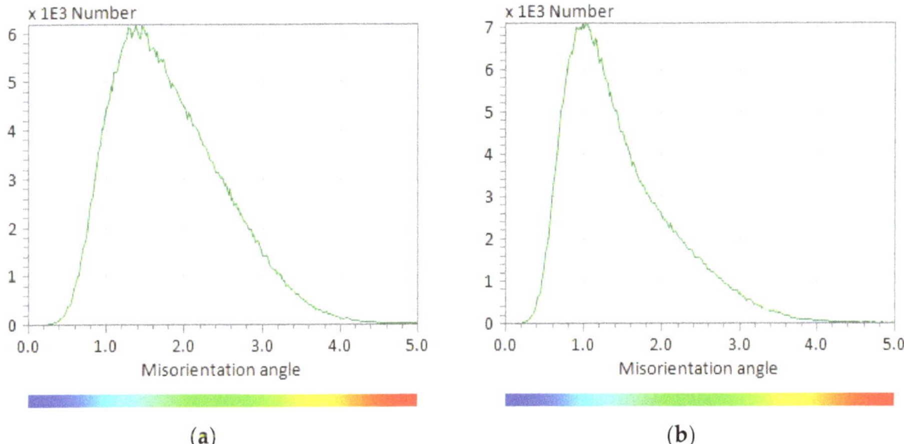

(**a**)

(**b**)

**Figure 10.** KAM changes on IC CGHAZ area during the four days of monitoring: (**a**) first day of monitoring and (**b**) fourth day of monitoring.

### 3.4.2. Redistribution of GAM

Grain average misorientation (GAM) maps show the correlated mean misorientation values obtained in a specific grain [44,45] and the degree of intragranular orientation deviation [7]. This means that each measure contained in a grain is assigned the same local misorientation value, but the values vary from one grain to another [16]. GAM excludes some external disturbances and manifests the global orientation deviations within a specific EBSD observational region [46]. This is a typical way to show orientation changes within the grain. Figures 11 and 12 show the results obtained by GAM, and a change in the fraction of the misorientation values can be observed from the first to the fourth day of observation. In Figure 11a, it can be seen in the red box that misorientation angles between values of 12 to 15° dropped to values of 5 to 10° in Figure 11b (according to the GAM color pattern). In Figure 12, we can see that the fraction of GAM values for 2° increased from the first to the fourth day and from 4° the fraction of GAM values decreased from the first to the last day of observation. As mentioned above, these changes in misorientation are related to a change in the concentration of the dislocations over time.

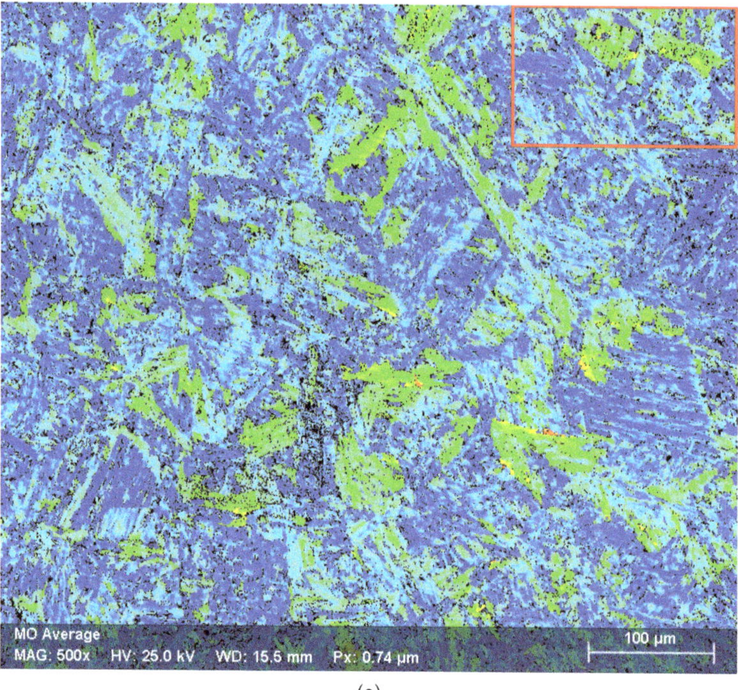

(a)

**Figure 11.** *Cont.*

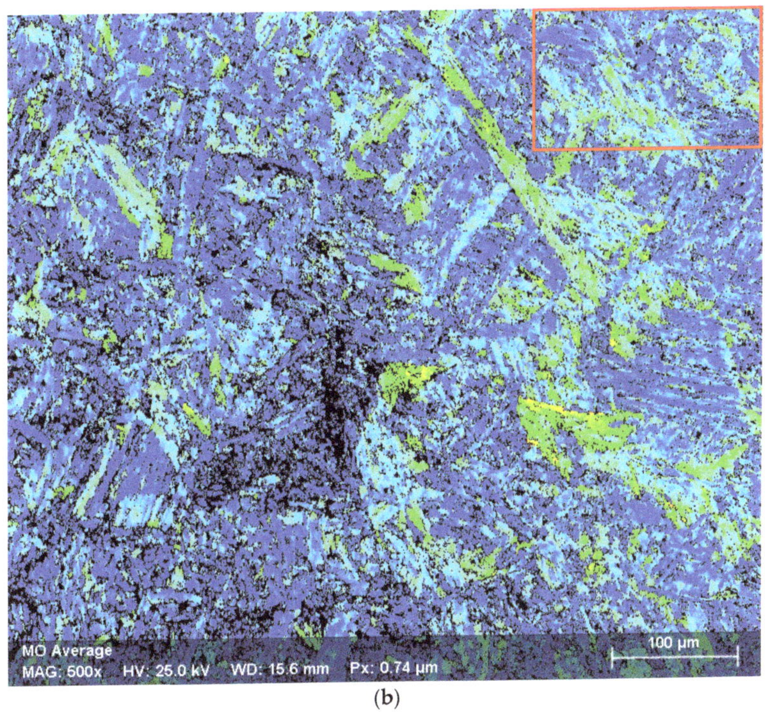

**Figure 11.** Grain average misorientation (GAM): (**a**) first day of monitoring GAM map and (**b**) fourth day of monitoring GAM map. EBSD 500×.

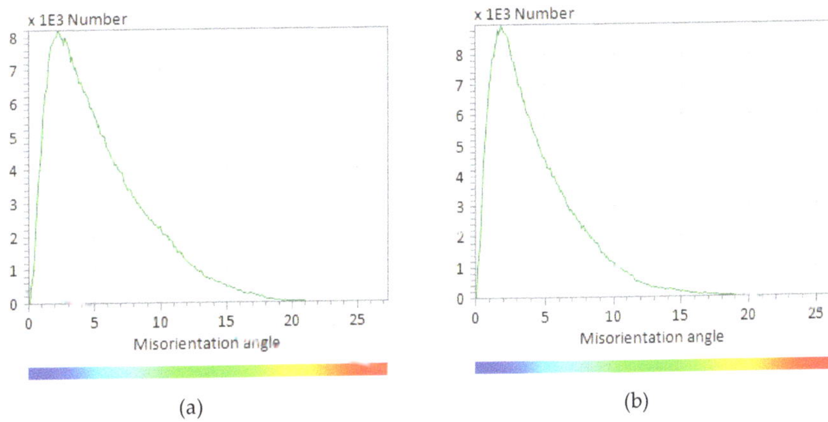

**Figure 12.** GAM changes on IC CGHAZ area during the four days of monitoring: (**a**) first day of monitoring and (**b**) fourth day of monitoring.

### 3.4.3. Redistribution of Grain Boundaries

Figure 13 shows the high-angle grain boundaries (HAGBs) in yellow and the low-angle grain boundaries (LAGBs) in blue. Figure 13b shows a greater fraction of the HAGB values > 50° than the fraction of LAGB <15°. The percentage of LAGBs that consist of a series of dislocations decreases at ICHAZ with an increase for the cooling elapsed time, which corresponds to high heat input [22]. The simulated samples of ICHAZ showed a selective orientation of misorientation and an apparently 'bimodal' feature [47–49]) in the range of 2–15° and 50–62°. The authors of [50] stated that the LAGBs usually distribute between the parallel or thinner grain laths and the HAGBs locate within the prior austenite grain boundaries. It can be determined that the preferred orientation phenomenon occurred during the phase transformation in a different or the same bainite group, which could generate HAGBs or LAGBs at the boundaries of laths, respectively [22]. HAGBs are more mobile than LAGBs due to a higher density of dislocations and a large amount of stored energy [31]. According to Tsuchiyama et al. [51], HAGBs show mobility and a driving force sufficiently large for migration. LAGBs are mobile because their properties depend on the density of dislocations [31].

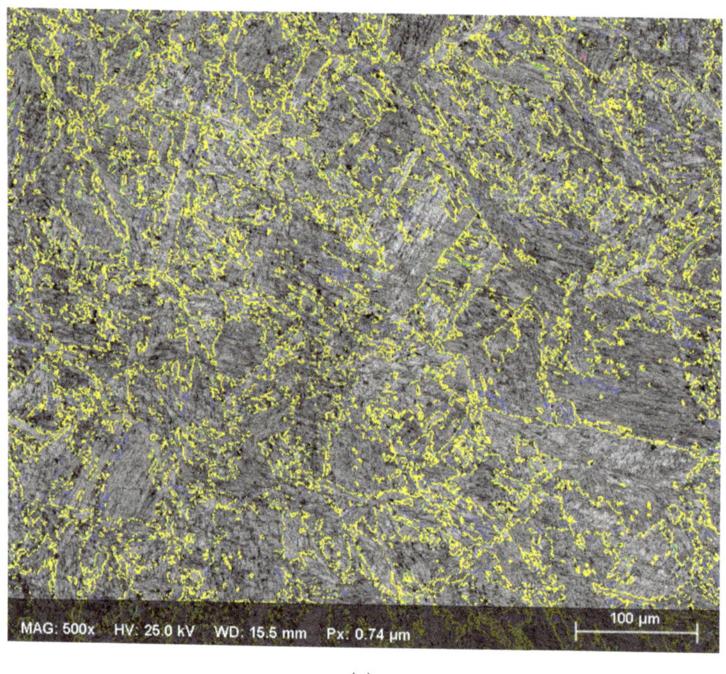


(**a**)

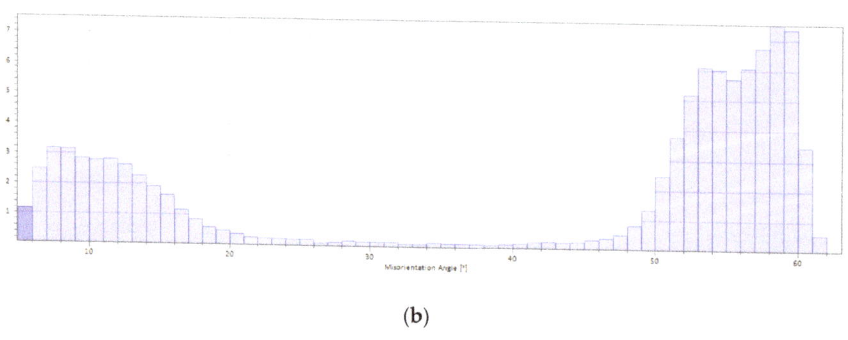


(**b**)

**Figure 13.** (**a**) Maps of boundary grain and (**b**) fraction of misorientation of boundaries grain.

## 4. Discussion

### *4.1. Microstructure of Base Metal and IC CGHAZ*

Figure 6, obtained with the EBSD technique, shows us the typical microstructure of a DH36 high-strength steel without physical simulation, which is mainly made up of ferrite accompanied by phases of superior bainite, austenite, and cementite. The black colored zero solutions in Figure 6b,c may represent the martensite of the MA microconstituent, which are smaller than 1 μm in size. In addition to the fact that EBSD may not have indexed microstructures smaller than the step size placed before indexing, microstructures such as martensite that are not present in the EBSD database are also not indexed. Non-indexed regions (represented as dark areas) contain a high density of dislocations [52]. Martensite identification is complicated with EBSD, due to the similar lattice structures of ferrite and martensite [53]. Wright [54] mentions that martensite cannot be indexed using EBSD as martensite has a body centered tetragonal (bct) structure but is only slightly tetragonal. Thus, EBSD cannot reliably identify this tetragonality. The c/a ratio is simply very close to 1, so no information was entered in the data bank to identify martensite. Nowell et al. [55] usually perform this indexing as a body centered cubic (bcc) ferrite structure and then use various approximations to try to differentiate martensite from ferrite based on topographic arguments.

When the two thermal welding cycles were physically simulated, the EBSD technique still detected phases belonging to the ferrite, upper bainite, austenite, and cementite in the same way as was detected for the base material. The difference between these two microstructures is evident: the banding presented in the base material disappeared, the grain size increased, and the distribution of the bainite, cementite, and austenite phases within the ferritic matrix varied (all this can be observed by comparing Figures 6 and 7). In the intercritically reheated coarse grain heat affected zone of an EH36 weld, Tsay et al. [56] observed microstructures such as lower bainite with carbides between those of upper bainite and with retained austenite. In the intercritically reheated heat affected zone of DH36 presented in slab bainite, this type of bainite generally nucleates along the grain boundary of the original austenite and then grows into the interior grain [22]. When the maximum heating temperature is between AC3 and AC1, the superheated austenite decomposes into Widmanstatten ferrite, bainite, and microphases consisting of martensite and retained austenite (MA) [57]. When the cooling time $t_{8/5}$ is increased, the formation of MA is more pronounced, and due to the semi-diffusive transformation of bainite and a mixture of ferrite and carbide, the formation of granular bainite plays a role in separating slab bainite [22]. Hutchinson et al. [58] obtained, by physical simulation, large grains of polygonal ferrite outlining the contours of the initial austenitic grain and bainitic microstructures with fine cementite particles.

### *4.2. Redistribution of the Residual Tension vs. the Misorientation*

The study of the variation of the crystallographic orientation and the local strain gradient for the prediction of stress relief is crucial [38]. Many studies have shown a good correlation between the degree of misorientation and macroscopic strain in the material [59]. The results of our research work have shown a variation of the residual welding stress (Figure 8) obtained with X-ray diffraction and a variation of the misorientation of grain contours (Figures 9–12) obtained with the methods KAM and GAM. Since misorientation relates microstructural deformation to high- and low-angle grain boundary dislocations, and residual stresses are the result of the concentration of dislocations within different regions of the microstructure, there is an obvious relationship for this variation obtained with the experimental methods used, due to the rearrangement of the dislocations after immediately having completed the welding process and through time where the stabilizing of residual stresses occurs, as obtained in the works of Estefen [8] and Gurova [9,10].

According to studies realized by Wright et al. [16], the residual deformation is manifested as a variation in the orientation of the network, and Han et al. [60] indicate that with increasing strain there is an increase in the density of low-angle grain boundaries.

Therefore, local misorientations provide an indication of the residual strain distribution in polycrystalline materials and for their characterization KAM-based methods are suitable [16]. Therefore, the KAM is often used as a qualitative measure of plastic deformation localization [25]. Orientation gradients can be correlated with geometrically necessary dislocations (GND) [61], which are generated to accommodate local incompatibilities within crystallographic parameters [16,62] and orientation gradients which are related to plastic deformation [63] and to correspond to the applied macroscopic strain when averaged over multiple grains or entire measurement fields [64]. The development of misorientation within the grain can be attributed to the high density of dislocations [65]. Regions with high concentrations of LAGB correspond to areas with a high concentration of dislocations [66]. The KAMs obtained in Figure 10 show that a large concentration of LAGBs was in the range of 1 to 2° of misorientation and over time the dislocations moved and changed the internal strain. The results indicate that the glide of the dislocations gives way to the reorganization of the LAGBs and HAGBs. When the deformation progresses, the misorientation across a specific sub-structural boundary will increase [25]. Thus, the initial dislocation cell structure created by the low-angle dense dislocation wall will evolve into sub-grain boundaries and eventually into new high-angle grain boundaries [25]; the HAGBs are effective barriers for dislocation motion [67]. Experimental results illustrated high KAM values near grain boundaries at lower strain levels, while a wider distribution in KAM values outside the grain boundaries was observed in larger strains [62]. At high strains, due to highly misoriented grains and the formation of submicron dislocation structures, it is crucial to choose the correct step size [62]. For the four-day observation period inside the SEM chamber, the same step size was used, ruling out that this had influenced the change in the misorientation values of the thermomechanically simulated microstructure.

In the GAM results, large changes in principle were found in the LAGBs from 2° to 10°, and as indicated in the literature, these changes in misorientation are related to a change in the concentration of the dislocations, in this case, over time. As great dislocations densities develop within the material, the residual strain is represented as local variations in the lattice orientation for GAM [68]. GAM maps for misorientations between 0° and 2° can be employed to estimate the internal energy or dislocation density, qualitatively, where the higher GAM values show higher dislocation densities [69]. According to Hou et al. [70], the regions close to the fusion zone, such as coarse-grained ones, present high GAM values. Recrystallization is related to LAGBs, and these values can be obtained using GAM maps [71]. The GAM evolution for the deformed material showed higher values than the initial annealed condition because of the significant number of dislocations [68]. When the deformation occurs and progressively increases, some grains with lower GAM values are distributed around grains with higher values [68]. When the GAM values obtained over many grains of a scanned area are averaged out, they can be correlated with macroscopic values of plastic deformation [72]. The decrease in misorientation obtained with GAM and the residual stresses clearly shows the effect of the role of the microstructure on the mechanical behavior of the welded joints [45]. Research related to GAM indicates that these values are influenced by the step size [72]; for the purposes of the research carried out, the step size remained constant during the 4 days of observation.

## 5. Conclusions

It was observed that there was a redistribution of residual stresses, as well as a redistribution of misorientation in the IC CGHAZ area, four days after having carried out the thermomechanical simulation. The main results can be summarized as follows:

(a) For the residual stresses, a decrease in residual compression stresses in the range of 196 MPa to 160 MPa was noted.

(b) For 1° of misorientation in KAM, on the first day the fraction values were 4% and on the fourth day the fraction values increased to 7%.

(c) For the misorientation in GAM, the fraction values for 2° increased from the first to the fourth day and for 4° the fraction values decreased from the first to the fourth day.

(d)     The high-angle grain contours and the low-angle grain contours showed rearrangement due to a rearrangement of the dislocations over time; these changes were related to the change in residual stresses that, after two weeks, stabilized.

**Author Contributions:** The first draft of the paper was written by G.S.C., the analysis of the final results in X-ray diffraction was done by A.L. and T.G., analysis of the equipment of the thermomechanical simulation, SEM, and EBSD was done by L.S.G. and S.B.P., and the final editing was done by S.F.E. and G.S.C. All authors have read and agreed to the published version of the manuscript.

**Funding:** The authors acknowledge the funding from scholarship PRH03 Petrobras for G.S.C., the Brazilian Research Council (CNPq)—Research Projects 456319/2013-1, PQ 305657/2017-8 funded to S.F.E. and Scholarship n° 124550/2018-5 to T.G., and the funding from the Fundação de Amparo à Pesquisa do Estado do Rio de Janeiro (FAPERJ) to T.G., and the funding from E-26/202.600/2019 (246899) to S.F.E.

**Data Availability Statement:** Not applicable.

**Acknowledgments:** The authors are grateful to Stuart I. Wright from EDAX-TSL Company for his comments on microstructure analysis using EBSD techniques. Special thanks to the SENAI Institute of Rio de Janeiro and the GURTEQ Company for providing the use of GLEEBLE, SEM-EBSD, and RAYSTRESS to perform this research work.

**Conflicts of Interest:** The authors declare no conflict of interest.

# References


1.  Masubuchi, K. *Analysis of Welded Structures Residual Stresses, Distortion, and Their Consequences*, 1st ed.; Pergamon Press: Cambridge, UK, 1980; p. 3. [CrossRef]
2.  Pillai, R.; Chyrkin, A.; Quadakkers, W.J. Modeling in High Temperature Corrosion: A Review and Outlook. *Oxid. Met.* **2021**, *96*, 385–436. [CrossRef]
3.  Cruz, V.; Chao, Q.; Birbilis, N.; Fabijanic, D.; Hodgson, P.; Thomas, S. Electrochemical studies on the effect of residual stress on the corrosion of 316L manufactured by selective laser melting. *Corros. Sci.* **2020**, *164*, 108314. [CrossRef]
4.  Chai, D.; Ma, G.; Zhou, S.; Jin, Z.; Wu, D. Cavitation erosion behavior of Hastelloy™ C-276 weld by laser welding. *Wear* **2019**, *420*, 226–234. [CrossRef]
5.  Bi, Y.; Yuan, X.; Lv, J.; Bashir, R.; Wang, S.; Xue, H. Effect of Yield Strength Distribution Welded Joint on Crack Propagation Path and Crack Mechanical Tip Field. *Materials* **2021**, *14*, 4947. [CrossRef]
6.  Nishimura, R.; Ma, N.; Liu, Y.; Li, W.; Yasuki, T. Measurement and analysis of welding deformation and residual stress in CMT welded lap joints of 1180 MPa steel sheets. *J. Manuf. Process.* **2021**, *72*, 515–528. [CrossRef]
7.  Wang, Y.; Feng, G.; Pu, X.; Deng, D. Influence of welding sequence on residual stress distribution and deformation in Q345 steel H-section butt-welded joint. *J. Mater. Sci. Technol.* **2021**, *13*, 144–153. [CrossRef]
8.  Estefen, S.F.; Gurova, T.; Werneck, D.; Leontiev, A. Welding stress relaxation effect in butt-jointed steel plates. *Mar. Struct.* **2012**, *29*, 211–225. [CrossRef]
9.  Gurova, T.; Estefen, S.F.; Leontiev, A.; Oliveira, F.A.L. Welding residual stresses: A daily history. *Sci. Technol. Weld. Join.* **2015**, *20*, 616–621. [CrossRef]
10. Gurova, T.; Estefen, S.F.; Leontiev, A.; Barbosa, P.T.; Oliveira, F.A.L. Time-dependent redistribution behavior of residual stress after repair welding. *Weld. World* **2017**, *61*, 507–515. [CrossRef]
11. Taylor, G.I. The mechanism of plastic deformation of crystals. Part II. Comparison with observations. *Proc. Roy. Soc. A* **1984**, *145*, 388–404. [CrossRef]
12. Orowan, E. Problems of Plastic Gliding. *Proc. Phys. Soc.* **1940**, *52*, 8–22. Available online: http://iopscience.iop.org/0959-5309/52/1/303 (accessed on 23 April 2021). [CrossRef]
13. Love, A.E.H. *A Treatise on the Mathematical Theory of Elasticity*, 1st ed.; Cambridge: Cambridge, UK, 1892.
14. François, D.; Pineau, A.; Zaoui, A. *Mechanical Behaviour of Materials, Micro and Macroscopic Constitutive Behaviour*, 1st ed.; Springer: Dordrecht, The Netherlands, 2012. [CrossRef]
15. Fan, H.; Wang, Q.; El-Awady, J.A.; Raabe, D.; Zaiser, M. Strain rate dependency of dislocation plasticity. *Nat. Commun.* **2021**, *12*, 1845. [CrossRef] [PubMed]
16. Wright, S.I.; Nowell, M.M.; Field, D.P. A review of strain analysis using electron backscatter diffraction. *Microsc. Microanal.* **2011**, *17*, 316–329. [CrossRef] [PubMed]
17. Merriman, C.C.; Field, D.P.; Trivedi, P. Orientation dependence of dislocation structure evolution during cold rolling of aluminum. *Mater. Sci. Eng. A* **2008**, *494*, 28–35. [CrossRef]
18. Kundu, A.; Field, D.P. Geometrically Necessary Dislocation Density Evolution in Interstitial Free Steel at Small Plastic Strains. *Metall. Mater. Trans. A* **2018**, *49*, 3274–3282. [CrossRef]

19.  Furuhara, T.; Chiba, T.; Kaneshita, T.; Wu, H.; Miyamoto, G. Crystallography and Interphase Boundary of Martensite and Bainite in Steels. *Metall. Mater. Trans. A* **2017**, *48*, 2739–2752. [CrossRef]

20.  Filho, I.S.; Sandim, M.J.R.; Ponge, D.; Sandim, H.; Raabe, D. Strain hardening mechanisms during cold rolling of a high-Mn steel: Interplay between submicron defects and microtexture. *Mater. Sci. Eng. A* **2019**, *754*, 636–649. [CrossRef]

21.  Tomota, Y.; Ojima, M.; Harjo, S.; Gong, W.; Sato, S.; Ungár, T. Dislocation densities and intergranular stresses of plastically deformed austenitic steels. *Mater. Sci. Eng.* **2019**, *743*, 32–39. [CrossRef]

22.  Zhang, S.; Liu, W.; Wan, J.; Misra, R.; Wang, Q.; Wang, C. The grain size and orientation dependence of geometrically necessary dislocations in polycrystalline aluminum during monotonic deformation: Relationship to mechanical behavior. *Mater. Sci. Eng. A* **2020**, *775*, 138939. [CrossRef]

23.  Shawish, S.E.; Cizelj, L.; Simonovski, I. Misorientation Effects in an Anisotropic Plasticity Finite Element Model of a Polycrystalline under Tensile Loading. In Proceedings of the 21st International Conference Nuclear Energy for New Europe, Ljubljana, Slovenia, 5–7 September 2012; Nuclear Society of Slovenia: Ljubljana, Slovenia, 2012; pp. 602.1–602.8.

24.  Porter, D.A.; Easterling, K.E.; Sherif, M.Y. *Phase Transformations in Metals and Alloys*, 3rd ed.; CRC Press Book: Boca Raton, NY, USA, 2009; pp. 122–126. ISBN 9781420062106.

25.  Lehto, P. Adaptive domain misorientation approach for the EBSD measurement of deformation induced dislocation sub-structures. *Ultramicroscopy* **2021**, *222*, 113203. [CrossRef]

26.  Li, J.-S.; Cheng, G.-J.; Yen, H.-W.; Yang, Y.-L.; Chang, H.-Y.; Wu, C.-Y.; Wang, S.-H.; Yang, J.-R. Microstrain and boundary misorientation evolution for recrystallized super DSS after deformation. *Mater. Chem. Phys.* **2020**, *246*, 122815. [CrossRef]

27.  Nolze, G.; Winkelmann, A. About the reliability of EBSD measurements: Data enhancement. In *IOP Conference Series: Materials Science and Engineering*; IOP Publishing: Bristol, UK, 2020; Volume 891, p. 012018.

28.  Tong, V.S.; Britton, T.B. TrueEBSD: Correcting spatial distortions in electron backscatter diffraction maps. *Ultramicroscopy* **2021**, *221*, 113130. [CrossRef]

29.  Zhang, X.; Lu, S.; Zhang, B.; Tian, X.; Kan, Q.; Kang, G. Dislocation–grain boundary interaction-based discrete dislocation dynamics modeling and its application to bicrystals with different misorientations. *Acta Mater.* **2021**, *202*, 88–98. [CrossRef]

30.  Costa, A.D.S.B.; Abreu, H.F.G.; Miranda, H.; Costa, R.C.S.; Teixeira, L.; Philipov, S. Comparing between residual stresses and grain boundary engineering for tubulations used in hydrodessulfuration systems. In Proceedings of the 4th Congress on P&D Oil and Gas, Campinas, São Paulo, Brazil, 21–24 September 2007; 4o PDPETRO: Campinas, Brazil, 2007; pp. 1–10.

31.  Humphreys, J.; Rohrer, G.S.; Rollett, A. *Recrystallization and Related Annealing Phenomena*, 3rd ed.; Elsevier: Amsterdam, The Netherlands, 2017; ISBN 978-0-08-098235-9.

32.  Vervynckt, S.; Verbeken, K.; Lopez, B.; Jonas, J.J. Modern HSLA steels and role of non-recrystallisation temperature. *Int. Mater. Rev.* **2012**, *57*, 187–207. [CrossRef]

33.  Chavez, G.F.S. Physical Simulation and Characterization on Heat Affected Zones of API 5L Grade x80 Steels. Master's Thesis, Escola Politecnica da Universidade de São Paulo, São Paulo, Brazil, 7 December 2011. [CrossRef]

34.  Moeinifar, S.S.; Kokabi, A.H.; Madaah Hosseini, H.R. Role of Tandem Submerged Arc Welding Thermal Cycles on Properties of the Heat Affected Zone in X80 Microalloyed Pipe Line Steel. *J. Mater. Process. Technol.* **2011**, *211*, 368–375. [CrossRef]

35.  Davis, C.L.; King, J.E. Cleavage Initiation in the Intercritically Reheated Coarse Grained Heat Affected Zone: Part II. *Metall. Mater. Trans. A* **1996**, *27*, 3019–3029. [CrossRef]

36.  You, Y.; Shanga, C.; Chen, L.; Subramanian, S. Investigation on the crystallography of the transformation products of reverted austenite in intercritically reheated coarse grained heat affected zone. *Mater. Des.* **2013**, *43*, 485–491. [CrossRef]

37.  Colpaert, H. *Metallography of Steels: Interpretation of Structure and the Effects of Processing*, 1st ed.; ASM International: Materials Park, OH, USA, 2018; ISBN 978-1-62708-148-1.

38.  Masoumi, M.; Echeverri, E.A.A.; Tschiptschin, A.P.; Goldenstein, H. Improvement of wear resistance in a pearlitic rail steel via quenching and partitioning processing. *Sci. Rep.* **2019**, *9*, 7454. [CrossRef] [PubMed]

39.  Zhang, Y.; Sabbaghianrad, S.; Yang, H.; Topping, T.D.; Langdon, T.G.; Lavernia, E.J.; Schoenung, J.; Nutt, S.R. Two-Step SPD Processing of a Trimodal Al-Based Nano-Composite. *Metall. Mater. Trans. A* **2015**, *46*, 5877–5886. [CrossRef]

40.  Jayashree, P.K.; Basu, R.; Sharma, S.S. An electron backscattered diffraction (EBSD) approach to study the role of microstructure on the mechanical behavior of welded joints in aluminum metal matrix composites. *Mater. Today Proc.* **2021**, *38*, 490–493. [CrossRef]

41.  Badji, R.; Chauveau, T.; Bacroix, B. Texture, misorientation and mechanical anisotropy in a deformed dual pase stainless steel weld joint. *Appl. Mater. Sci. Eng. A* **2013**, *575*, 94–103. [CrossRef]

42.  Schayes, C.; Bouquerel, J.; Vogt, J.-B.; Palleschi, F.; Zaefferer, S. A comparison of EBSD based strain indicators for the study of Fe-3Si steel subjected to cyclic loading. *Mater. Charact.* **2016**, *115*, 61–70. [CrossRef]

43.  Mohtadi-Bonab, M.A.; Eskandari, M.; Szpunar, J.A. Texture, local misorientation, grain boundary and recrystallization fraction in pipeline steels related to hydrogen induced cracking. *Mater. Sci. Eng. A* **2015**, *620*, 97–106. [CrossRef]

44.  Wright, S.I. Quantification of recrystallized fraction from orientation imaging scan. In Proceedings of the 20th International Conference on Textures of Materials, Montreal, QC, Canada, 9–13 August 1999; Szpunar, J.A., Ed.; NRC Research Press: Otawa, ON, Canada, 1999; pp. 104–109.

45.  Allain-Bonasso, N.; Wagner, F.; Berbenni, S.; Field, D.P. A study of the heterogeneity of plastic deformation in IF steel by EBSD. *Mater. Sci. Eng. A* **2012**, *548*, 56–63. [CrossRef]

46. Rui, S.-S.; Shang, Y.-B.; Fan, Y.-N.; Han, Q.-N.; Niu, L.; Shi, H.-J.; Hashimoto, K.; Komai, N. EBSD analysis of creep deformation induced grain lattice distortion: A new method for creep damage evaluation of austenitic stainless steels. *Mater. Sci. Eng. A* **2018**, *733*, 329–337. [CrossRef]

47. Pandey, C.; Mahapatra, M.; Kumar, P.; Saini, N.; Srivastava, A. Microstructure and mechanical property relationship for different heat treatment and hydrogen level in multi-pass welded P91 steel joint. *J. Manuf. Process.* **2017**, *28*, 220–234. [CrossRef]

48. Pandey, C.; Mahapatra, M.M.; Kumara, P.; Saini, N. Some studies on P91 steel and their weldments. *J. Alloys Compd.* **2018**, *743*, 332–364. [CrossRef]

49. Pandey, C.; Mahapatra, M.M.; Kumar, P.; Thakre, J.; Saini, N. Role of evolving microstructure on the mechanical behaviour of P92 steel welded joint in as-welded and post weld heat treated state. *J. Mater. Process. Tech.* **2019**, *263*, 241–255. [CrossRef]

50. Pereira, H.B.; Azevedo, C.R.F. Can the drop evaporation test evaluate the stress corrosion cracking susceptibility of the welded joints of duplex and super duplex stainless steels? *Eng. Fail. Anal.* **2019**, *99*, 235–247. [CrossRef]

51. Tsuchiyama, T.; Natori, M.; Nakada, N.; Takakis, S. Conditions for Grain Boundary Bulging during Tempering of Lath Martensite in Ultra-low Carbon Steel. *ISIJ Int.* **2010**, *50*, 771–773. [CrossRef]

52. Bozzolo, N.; Bernacki, M. Viewpoint on the Formation and Evolution of Annealing Twins during Thermomechanical Processing of FCC Metals and Alloys. *Met. Mater. Trans. A* **2020**, *51*, 2665–2684. [CrossRef]

53. Baghdadchi, A.; Vahid, A.; Karlsson, H.L. Identification and quantification of martensite in ferritic-austenitic stainless steels and welds. *J. Mater. Res. Tech.* **2021**, *15*, 3610–3621. [CrossRef]

54. Wright, S.I.; EDAX TSL, California USA. Personal communication, 2016.

55. Nowell, M.M.; Wright, S.I.; Carpenter, J.O. Differentiating ferrite and martensite in steel microstructures using electron backscatter diffraction. In Proceedings of the Materials Science and Technology (MS&T), Pittsburgh, PA, USA, 25–29 October 2009; Materials Science and Technology (MS&T): Pittsburgh, PA, USA, 2009; pp. 933–943.

56. Tsay, L.W.; Chern, T.S.; Gau, C.Y.; Yang, J.R. Microstructures and fatigue crack growth of EH36 TMCP steel weldments. *Int. J. Fatigue* **1999**, *21*, 857–864. [CrossRef]

57. Bhadeshia, H.K.D.H.; Svensson, L.-E.; Gretoft, B. A Model for the Development of Microstructure in Low-Alloy Steel (Fe-Mn-Si-C) Weld Deposits. *Acta Met.* **1985**, *33*, 1271–1283. [CrossRef]

58. Hutchinson, W.B.; Komenda, J.; Rohrer, G.S.; Beladic, H. Heat affected zone microstructures and their influence on toughness in two microalloyed HSLA steels. *Acta Mater.* **2015**, *97*, 380–391. [CrossRef]

59. Unnikrishnan, R.; Northover, S.M.; Jazaeri, H.; Bouchard, P.J. Investigating plastic deformation around a reheat-crack in a 316H austenitic stainless steel weldment by misorientation mapping. *Procedia Struct. Integr.* **2016**, *2*, 3501–3507. [CrossRef]

60. Han, J.H.; Jee, K.K.; Oh, K.H. Orientation rotation behavior during in situ tensile deformation of polycrystalline 1050 aluminum alloy. *Int. J. Mech. Sci.* **2003**, *45*, 1613–1623. [CrossRef]

61. Rui, S.-S.; Niu, L.-S.; Shi, H.-J.; Wei, S.; Tasan, C.C. Diffraction-based misorientation mapping: A continuum mechanics description. *J. Mech. Phys. Solids.* **2019**, *133*, 103709. [CrossRef]

62. Subedi, S.; Pokharel, R.; Rollet, A.D. Orientation gradients in relation to grain boundaries at varying strain level and spatial resolution. *Mater. Sci. Eng. A* **2015**, *638*, 348–356. [CrossRef]

63. Kamaya, M.; Wilkinson, A.J.; Titchmarsh, J.M. Measurement of plastic strain of polycrystalline material by electron backscatter diffraction. *Nucl. Eng. Des.* **2005**, *235*, 713–725. [CrossRef]

64. Kamaya, M. Assessment of local deformation using EBSD: Quantification of accuracy of measurement and definition of local gradient. *Ultramicroscopy* **2021**, *111*, 1189–1199. [CrossRef] [PubMed]

65. Jiang, J.; Britton, T.B.; Wilkinson, A.J. Measurement of geometrically necessary dislocation density with high resolution electron backscatter diffraction: Effects of detector binning and step size. *Ultramicroscopy* **2013**, *125*, 1–9. [CrossRef] [PubMed]

66. Gottstein, G.; Shvindlerman, L.S. *Grain Boundary Migration in Metals: Thermodynamics, Kinetics, Applications*, 2nd ed.; Taylor and Francis Group: Boca Raton, NY, USA, 2010. [CrossRef]

67. Wilson, D.; Wan, W.; Dunne, F.P.E. Microstructurally-sensitive fatigue crack growth in HCP, BCC and FCC polycrystals. *J. Mech. Phys. Solids.* **2019**, *126*, 204–225. [CrossRef]

68. Munoz, J.A.; Higucra, O.F.; Benito, J.A.; Bradai, D.; Khelfa, T.; Bolmaro, R.E.; Jorge, A.M.; Cabrera, J.M. Analysis of the micro and substructural evolution during severe plastic deformation of ARMCO iron and consequences in mechanical properties. *Mater. Sci. Eng. A* **2019**, *740*, 108–120. [CrossRef]

69. Heidarzadeh, A.; Radi, A.; Yapici, G. Formation of nano-sized compounds during friction stir welding of CueZn alloys: Effect of tool composition. *J. Mater. Res. Technol.* **2020**, *9*, 15874–15879. [CrossRef]

70. Hou, J.; Shoji, T.; Lu, Z.; Peng, Q.; Wang, J.; Han, E.-H.; Ke, W. Residual strain measurement and grain boundary characterization in the heat-affected zone of a weld joint between alloy 690TT and alloy 52. *J. Nucl. Mater.* **2010**, *397*, 109–115. [CrossRef]

71. Mirzadeh, H.; Cabrera, J.M.; Najafizadeh, A.; Calvillo, P.R. EBSD study of a hot deformed austenitic stainless steel. *Mater. Sci. Eng. A* **2012**, *538*, 236–245. [CrossRef]

72. Guglielmi, P.O.; Ziehmer, M.; Lilleoddena, E.T. On a novel strain indicator based on uncorrelated misorientation angles for correlating dislocation density to local strength. *Acta Mater.* **2018**, *150*, 195–205. [CrossRef]

*Article*

# Cold Formabilities of Martensite-Type Medium Mn Steel

Koh-ichi Sugimoto [1],*, Hikaru Tanino [1] and Junya Kobayashi [2]

[1] School of Science and Technology, Shinshu University, Nagano 380-8553, Japan; 14tm116j@shinshu-u.ac.jp
[2] School of Science and Engineering, Ibaraki University, Hitachi 316-8551, Japan; junya.kobayashi.jkoba@vc.ibaraki.ac.jp
* Correspondence: sugimot@shinshu-u.ac.jp; Tel.: +81-90-9667-4482

**Abstract:** Cold stretch-formability and stretch-flangeability of 0.2%C-1.5%Si-5.0%Mn (in mass%) martensite-type medium Mn steel were investigated for automotive applications. High stretch-formability and stretch-flangeability were obtained in the steel subjected to an isothermal transformation process at temperatures between $M_s$ and $M_f - 100\ °C$. Both formabilities of the steel decreased compared with those of 0.2%C-1.5%Si-1.5Mn and -3Mn steels (equivalent to TRIP-aided martensitic steels), despite a larger or the same uniform and total elongations, especially in the stretch-flangeability. The decreases were mainly caused by the presence of a large amount of martensite/austenite phase, although a large amount of metastable retained austenite made a positive contribution to the formabilities. High Mn content contributed to increasing the stretch-formability.

**Keywords:** stretch-formability; stretch-flangeability; martensite-type medium Mn steel; retained austenite; heat-treatment; isothermal transformation process; direct quenching

**Citation:** Sugimoto, K.-i.; Tanino, H.; Kobayashi, J. Cold Formabilities of Martensite-Type Medium Mn Steel. *Metals* **2021**, *11*, 1371. https://doi.org/10.3390/met11091371

Academic Editor: Beatriz López Soria

Received: 2 August 2021
Accepted: 27 August 2021
Published: 30 August 2021



## 1. Introduction

To date, the first, second and third-generation advanced ultrahigh- and high-strength sheet steels (AHSSs) have been developed for the weight reduction and high crush safety of automobiles [1–4]. In most AHSSs, their ductility is enhanced by transformation-induced plasticity (TRIP) [5] and/or twinning-induced plasticity (TWIP) [6] of metastable retained austenite, reverted austenite, and austenite. These AHSSs are categorized as follows [4],

1. First-generation AHSS: ferrite-martensite dual-phase (DP) steel [2,3], TRIP-aided polygonal ferrite (TPF) steel [2], TRIP-aided annealed martensite (TAM) steel [7] and complex-phase (CP) steel [3],
2. Second-generation AHSS: high-Mn TWIP and TWIP/TRIP steels [6],
3. Third-generation AHSS (Type A): TRIP-aided bainitic ferrite (TBF) steel [3,8–10], one-step and two-step quenching and partitioning (Q&P) steels [3,11–13], carbide-free bainitic (CFB) steel [14–16], and duplex-type, laminate-type, and bainitic ferrite-type medium manganese (D–MMn [17–22], L-MMn [23–25], and BF-MMn [26]) steels,
4. Third-generation AHSS (Type B): TRIP-aided martensitic (TM) steel [27–29] and martensite-type medium manganese (M–MMn) steel [30–33].

The second-generation AHSSs are mainly high-Mn austenitic steels with an Mn content higher than 14 mass% and possess the highest ductility by TWIP/TRIP effects. The first- and third-generation AHSSs contain metastable retained austenite or reverted austenite of 5 to 40 vol.%—except for the DP and CP steels—and possess high ductility and cold formability by the TRIP effect of the retained austenite or reverted austenite. The third-generation AHSSs are classified into two types, Type A and Type B, by their kinds of matrix structures and tensile strengths (TS) [4]. The matrix structures and the TSs of Type A are bainitic ferrite (BF) and bainitic ferrite/martensite (BF/M), and higher than 1.0 GPa, respectively, except for D–MMn and L-MMn steels with an annealed martensite [17–25] and/or δ-ferrite [24] matrix structure. On the other hand, the main matrix structure of Type B is martensite and its TS is higher than 1.5 GPa [27–33].

The M–MMn steel contains a larger amount of retained austenite than the TM steel [30–33], although its amount is much less than those of the D–MMn and L-MMn steels [17–25]. Resultantly, the M–MMn steel has an ultra-high tensile strength and shows a large elongation by the TRIP effect of the retained austenite, although the impact toughness is inferior to that of TM steel [30]. If the M–Mn steel achieves higher cold formability than the TM and D–MMn steels, significant weight reduction of the automobiles is expected. Many researchers have reported the cold formability of D–MMn steel because the steel achieves superior formability [22]. However, the cold formability of the M–MMn steel has been hardly investigated to date, although its total elongation is higher than those of TM and D–MMn steels [31].

In this study, the ductility, stretch-formability, and stretch-flangeability of M–MMn steel with a chemical composition of 0.2%C-1.5%Si-5.0%Mn (in mass%) were investigated for applications to the cold-stamping and cold-forging of automotive parts. In addition, these formabilities were related to the microstructural properties. Moreover, the formabilities were compared to the steels with lower Mn content, which are equivalent to TM steels.

## 2. Experimental Procedures

A steel containing 5 mass% Mn was prepared in the form of 100 kg slabs by vacuum melting. To investigate the Mn-content dependencies of microstructure, tensile properties, and formability, the steel slabs, containing low Mn content (1.5 and 3 mass% Mn) were also prepared. Hereafter, the steels with 1.5, 3, and 5 mass% Mn are referred to as 1.5Mn, 3Mn, and 5Mn steels, respectively. The chemical composition, austenite-finish and -start temperatures ($Ac_3$, $Ac_1$ in °C), and martensite-start and -finish temperatures ($M_s$ and $M_f$ in °C) of these slabs are shown in Table 1. The slabs were then heated to 1200 °C and hot-rolled to 5 mm thickness with a finishing temperature of 850 °C, followed by air-cooling to room temperature. After being ground to a thickness of 3 mm, the plates were cold-rolled into sheets of 1.2 mm thickness, with the assistance of annealing, at 650 °C. The CCT diagrams of these steels are shown in Figure 1a. No bainitic transformation appears only in 5Mn steel in a cooling rate range above 0.3 °C/s. The 1.5Mn and 3Mn steels subjected to the DQ process and the IT process at temperatures lower than $M_f$ is equivalent to the TM steel [27–29].

**Table 1.** Chemical composition (mass%) and various transformation temperatures ($Ac_3$, $Ac_1$, $M_s$, and $M_f$ in °C) of steels used.

| Steel | C | Si | Mn | P | S | Al | N | O | $Ac_3$ | $Ac_1$ | $M_s$ | $M_f$ |
|---|---|---|---|---|---|---|---|---|---|---|---|---|
| 1.5Mn | 0.20 | 1.49 | 1.50 | 0.006 | 0.0015 | 0.035 | 0.0038 | <0.001 | 847 | 719 | 420 | 300 |
| 3Mn | 0.20 | 1.52 | 2.98 | 0.006 | 0.0016 | 0.037 | 0.0034 | <0.001 | 797 | 689 | 363 | 220 |
| 5Mn | 0.21 | 1.50 | 4.94 | 0.005 | 0.0016 | 0.032 | 0.0020 | <0.001 | 741 | 657 | 282 | 150 |

Tensile specimens with 50 mm gauge length and 12.5 mm width parallel to the rolling direction and stretch-forming and stretch-flanging specimens with dimensions of 50 mm square were machined from the cold-rolled steel sheets. These specimens were subjected to the heat treatment shown in Figure 1b, namely, direct quenching in oil at 25 °C (DQ) and isothermal transformation (IT) at $T_{IT}$ = 100 °C to 450 °C (below $M_f$ to above $M_s$) for $t_{IT}$ = 1 × 10² s to 1 × 10⁵ s after being austenitized at 800 °C to 900 °C for 1200 s. The IT times used are corresponding to the times for which the maximum retained austenite fraction is obtained. Hereafter, the DQ process is also dealt with as the IT process at $T_{IT}$ = 25 °C.

The microstructure of the steels was observed by field-emission scanning electron microscopy (FE-SEM; JSM-6500F, JEOL Ltd., Akishima, Tokyo, Japan), which was performed using an instrument equipped with an electron backscatter diffraction (EBSD; OIM system, TexSEM Laboratories, Inc., Prova, UT, USA) system. The EBSD analyses were conducted in an area of 40 μm × 40 μm with a beam diameter of 1.0 μm and a beam step size of 0.1 μm operated at an acceleration voltage of 25 kV. The specimens for the FE-SEM–EBSD analysis were first ground with alumina powder and colloidal silica, and then ion thinning was carried out. The volume fraction of carbide in the specimens was measured using carbon extraction replicas and FE-SEM. The volume fractions of fine martensite/austenite constituent (MA phase) and the carbide were quantified by the line-intersecting method, as well as by the prior austenitic grain size and void density and diameter.

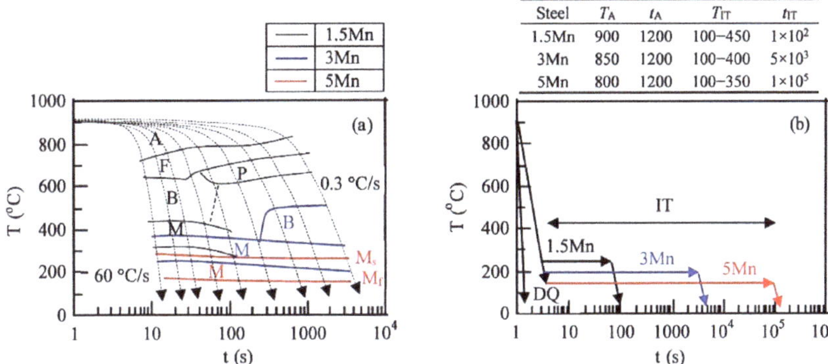



| Steel | $T_A$ | $t_A$ | $T_{IT}$ | $t_{IT}$ |
|-------|-------|-------|----------|----------|
| 1.5Mn | 900 | 1200 | 100–450 | $1\times10^2$ |
| 3Mn | 850 | 1200 | 100–400 | $5\times10^3$ |
| 5Mn | 800 | 1200 | 100–350 | $1\times10^5$ |

**Figure 1.** (**a**) CCT diagrams and (**b**) heat treatment diagram of the 1.5Mn, 3Mn, and 5Mn steels. "A", "F", "P", "B", and "M" in (**a**) are austenite, ferrite, pearlite, bainite, and martensite, respectively. "DQ" and "IT" represent direct quenching to room temperature and isothermal transformation, respectively. $T_A$ (°C): austenitizing temperature, $t_A$ (s): austenitizing time, $T_{IT}$ (°C): IT temperature, $t_{IT}$ (s): IT time.

The retained austenite characteristics of the steels were evaluated by an X-ray diffractometer (RINT2000, Rigaku Co., Akishima, Tokyo, Japan). The surfaces of the specimens were electropolished after being ground with emery paper (#1200). The volume fraction of the retained austenite phase ($f\gamma$, vol.%) was quantified from the integrated intensity of the (200)$\alpha$, (211)$\alpha$, (200)$\gamma$, (220)$\gamma$, and (311)$\gamma$ peaks obtained by X-ray diffractometry using Mo-K$\alpha$ radiation [34]. The carbon concentration ($C\gamma$, mass%) of the retained austenite was estimated from the empirical equation proposed by Dyson and Holmes [35]. To accomplish this, the lattice constant of retained austenite ($a_\gamma$, nm) was determined from the (200)$\gamma$, (220)$\gamma$, and (311)$\gamma$ peaks of the Cu-K$\alpha$ radiation. In this research, the average values of the volume fractions and carbon concentrations of retained austenite measured at three locations were adopted, as well as for other microstructural properties.

Tensile tests were carried out on a tensile-testing machine (AD-10TD, Shimadzu Co., Kyoto, Japan) at 25 °C, at a mean strain rate of $2.8 \times 10^{-3}$ s$^{-1}$ (crosshead speed: 10 mm/min). Stretch-forming tests were performed using the same testing machine used in the tensile tests to measure the maximum stretch height ($H_{max}$) at which a crack initiates. The forming temperature was 25 °C and the crosshead speed was 1 mm/min. A round punch tool with a curvature radius of 8.7 mm and a graphite lubricant were employed for the stretch-forming (Figure 2a). Hole-punching and hole-expanding tests were also conducted using the same testing machine to measure the hole-expansion ratio (HER) determined by

$$\text{HER} = (d_f - d_0)/d_0 \tag{1}$$

where $d_0$ and $d_f$ are the original hole diameter and the hole diameter upon cracking, respectively. First, a hole with a diameter of 4.76 mm was punched out at 25 °C at a punching rate of 10 mm/min, with a clearance of 10% between the die and punch using graphite lubricant (Figure 2b). Subsequently, hole expansion tests were performed using a conical punch tool with a vertical angle of 60 deg. at a crosshead rate of 1 mm/min, contacting with the roll-over section of the hole-punched specimen (Figure 2c). The hole-expanding tests were conducted at 25 °C, using a graphite lubricant. At least two specimens were tested for each condition to obtain the tensile, stretch-forming, and stretch-flanging properties.

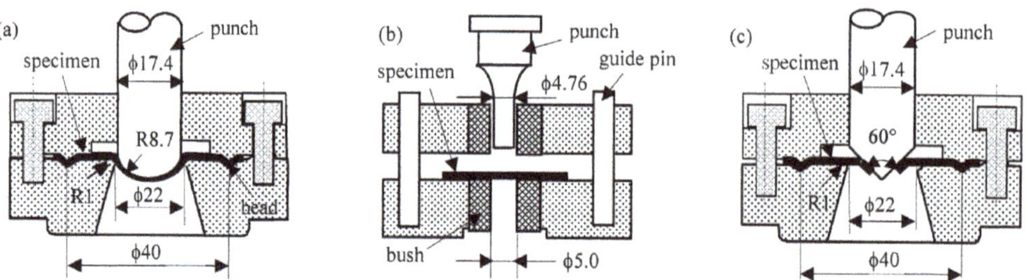


**Figure 2.** Die dimensions for (**a**) stretch-forming, (**b**) hole-punching and (**c**) hole-expanding tests.

### 3. Results

*3.1. Microstructure*

Figure 3 shows typical image quality distribution maps of the Fe-$\alpha$ (BCC) phase and FE-SEM images of carbon extraction replicas in the 1.5Mn, 3Mn, and 5Mn steels subjected to the IT process at the temperatures of $M_f$ + (30 °C to 70 °C). When the IT process was performed at the temperatures between $M_s$ and $M_f$, the microstructures of the 1.5Mn and 3Mn steels consisted of a mixture of coarse martensite and bainitic ferrite, MA phase, retained austenite, and carbide. In Figure 3a–c, the retained austenite phases look like black phases or points, because of unindexed phases or points. When the IT temperature was higher than $M_s$ and lower than $M_f$, their matrix structures changed into a single phase of bainitic ferrite or martensite, respectively. However, the matrix structure of the 5Mn steel changed into coarse martensite at all IT temperatures. In these steels, most retained austenites are mainly present in the MA phase and along the prior austenitic grain, packet, and block boundaries. The size of the coarse martensite and retained austenite phases tends to decrease with increasing Mn content. Most carbides precipitated only in coarse martensite. With increasing Mn content, both the volume fractions of the MA phase and carbide increases (see the bottom of Figure 3). The prior austenitic grain size slightly decreases with increasing Mn content, with a similar grain morphology.

Table 2 and Figure 4 show the initial volume fraction and carbon concentration of retained austenite and the strain-induced transformation factor ($k$) in the 1.5Mn, 3Mn, and 5Mn steels. The $k$ means the mechanical stability of retained austenite defined by the following equation [9],

$$k = (\log f\gamma_0 - \log f\gamma)/\varepsilon \qquad (2)$$

where $f\gamma_0$ is the initial volume fraction of retained austenite and $f\gamma$ is the retained austenite fraction after plastically strained to $\varepsilon$. The volume fractions of retained austenite in the 1.5Mn and 3Mn steels increase with increasing IT temperature. On the other hand, the volume fraction of the 5Mn steel becomes peak at an IT temperature between $M_s$ and $M_f$ and drastically decreases at the IT temperatures above $M_s$. When the initial retained austenite fractions of the steels are compared at an IT temperature range between $M_s$ and $M_f$, the retained austenite fraction increases with increasing Mn content. The maximum initial carbon concentrations of the steels are obtained at the IT temperatures between $M_s$ and $M_f$. The initial carbon concentration roughly decreases with increasing Mn content. This is considered to be caused by the stabilization effect of Mn enrichment in the retained austenite [21]. Although the IT temperature dependence of $k$-value is complex, the $k$-values of the 3Mn and 5Mn steels are lower than that of the 1.5Mn steel, except for the $k$-values at $T_{IT}$ > 350 °C in the 3Mn steel. The lower $k$-values of the 5Mn steel may be caused by that most of the retained austenites are finer than that of the 1.5Mn steel and surrounded by fine martensite.

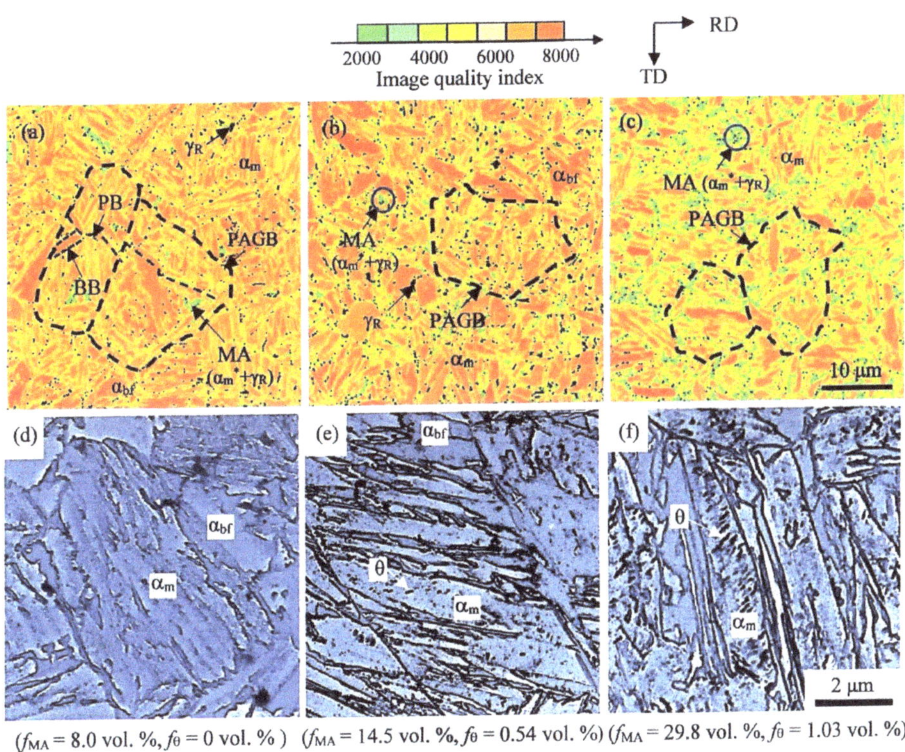



($f_{MA}$ = 8.0 vol. %, $f_\theta$ = 0 vol. % )  ($f_{MA}$ = 14.5 vol. %, $f_\theta$ = 0.54 vol. %)  ($f_{MA}$ = 29.8 vol. %, $f_\theta$ = 1.03 vol. %)

**Figure 3.** (**a–c**) Image-quality distribution maps of $\alpha$-Fe (BCC) phase and (**d–f**) FE-SEM images of replicas in the 1.5Mn (**a,d**), 3Mn (**b,e**), and 5Mn (**c,f**) steels subjected to the IT process at $T_{IT}$ = 370 °C, 280 °C, and 180 °C, respectively. The IT temperatures correspond to $M_f$ + (30 °C to 70 °C). PAGB, PB, and BB: prior austenitic grain, packet, and block boundaries, respectively. $\alpha_m$: coarse martensite, $\alpha_m^*$: fine martensite, $\alpha_{bf}$: bainitic ferrite, $\gamma_R$: retained austenite, $\theta$: carbide, MA: mixed phase of $\alpha_m^*$ and $\gamma_R$. $f_{MA}$ and $f_\theta$ are volume fractions of the MA phase and carbide, respectively. RD and TD represent rolling and thickness directions, respectively.

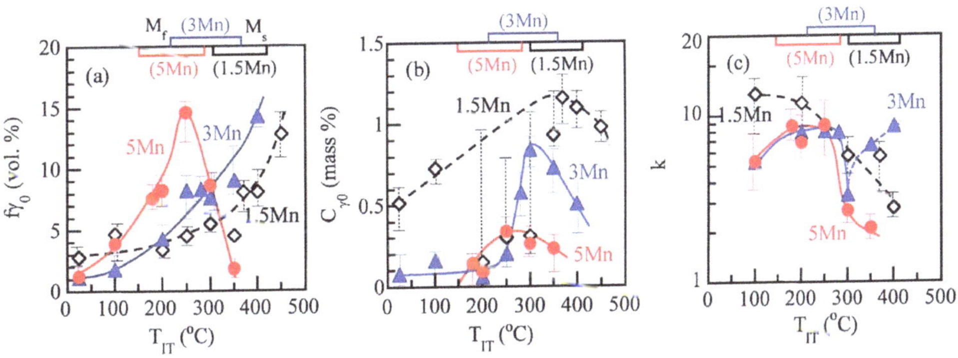



**Figure 4.** Variations in (**a**) initial volume fraction ($f\gamma_0$) and (**b**) carbon concentration ($C\gamma_0$) of retained austenite and (**c**) strain-induced transformation factor ($k$) as a function of isothermal transformation temperature ($T_{IT}$) in the 1.5Mn ($\diamond$), 3Mn ($\blacktriangle$), and 5Mn ($\bullet$) steels.

**Table 2.** Retained austenite characteristics, tensile properties, stretch-formability, and stretch-flangeability of the 1.5Mn, 3Mn, and 5Mn steels subjected to the IT process. Gray color zones correspond to IT temperatures between $M_s$ and $M_f$ of the steels. Dotted frames correspond to the IT temperatures between $M_s$ and $M_f - 100\,°C$.

| Steel | $T_{IT}$ | $f\gamma_0$ | $C\gamma_0$ | $k$ | YS | TS | UEl | TEl | RA | $H_{max}$ | TS × $H_{max}$ | HER | TS × HER |
|---|---|---|---|---|---|---|---|---|---|---|---|---|---|
| 1.5Mn | 25 | 2.8 | 0.52 | - | 1110 | 1527 | 4.0 | 8.7 | 40.3 | 7.73 | 11.8 | 27.4 | 41.8 |
|  | 100 | 4.7 | 0.73 | 13.5 | 940 | 1576 | 4.2 | 8.7 | 42.5 | 7.18 | 11.3 | 17.3 | 27.3 |
|  | 200 | 3.4 | 0.15 | 11.9 | 820 | 1580 | 5.3 | 9.9 | 42.4 | 8.33 | 13.2 | 37.6 | 59.4 |
|  | 250 | 4.5 | 0.30 | - | 960 | 1482 | 6.2 | 12.7 | 50.5 | 7.53 | 11.2 | 37.8 | 56.0 |
|  | 300 | 5.5 | 0.31 | 5.7 | 970 | 1438 | 4.3 | 10.4 | 48.0 | 8.70 | 12.5 | 42.8 | 61.5 |
|  | 350 | 4.5 | 0.93 | - | 960 | 1244 | 3.7 | 10.8 | 59.9 | 8.66 | 10.8 | 44.6 | 55.6 |
|  | 370 | 8.1 | 1.18 | 5.7 | 1025 | 1274 | 5.3 | 13.8 | 59.2 | 8.35 | 10.6 | 56.3 | 71.7 |
|  | 400 | 8.1 | 1.16 | 2.8 | 910 | 1111 | 4.2 | 12.1 | 65.4 | 9.23 | 10.3 | 64.9 | 72.2 |
|  | 450 | 12.8 | 1.10 | - | 690 | 887 | 16.2 | 24.8 | 63.7 | 9.43 | 8.4 | 59.5 | 52.8 |
| 3Mn | 25 | 1.1 | 0.08 | - | 1180 | 1656 | 4.7 | 4.7 | 1.0 | 7.93 | 13.1 | 30.7 | 50.8 |
|  | 100 | 1.8 | 0.16 | 5.3 | 1070 | 1603 | 6.5 | 14.1 | 54.5 | 7.92 | 12.7 | 26.1 | 41.9 |
|  | 200 | 4.3 | 0.06 | 8.0 | 970 | 1546 | 8.1 | 16.0 | 51.5 | 8.20 | 12.7 | 32.9 | 50.8 |
|  | 250 | 8.2 | 0.20 | 8.0 | 940 | 1481 | 7.2 | 14.5 | 53.1 | 8.45 | 12.5 | 28.8 | 42.6 |
|  | 280 | 8.3 | 0.58 | 8.0 | 1035 | 1375 | 6.9 | 14.6 | 56.5 | 8.39 | 11.5 | 30.3 | 41.7 |
|  | 300 | 7.6 | 0.84 | 3.3 | 1060 | 1345 | 5.7 | 13.6 | 56.7 | 8.43 | 11.3 | 37.6 | 50.6 |
|  | 350 | 9.0 | 0.73 | 6.7 | 930 | 1268 | 10.8 | 17.7 | 50.3 | 8.39 | 10.6 | 33.4 | 42.3 |
|  | 400 | 14.3 | 0.51 | 8.6 | 700 | 1265 | 13.3 | 18.5 | 39.2 | 7.23 | 9.1 | 1.6 | 2.0 |
| 5Mn | 25 | 1.2 | - | - | 1230 | 1907 | 5.3 | 5.3 | 5.0 | 2.40 | 4.6 | 2.7 | 5.2 |
|  | 100 | 3.9 | - | 5.3 | 930 | 1757 | 7.7 | 15.4 | 54.2 | 6.26 | 11.0 | 16.7 | 29.3 |
|  | 180 | 7.6 | 0.14 | 8.7 | 968 | 1633 | 7.8 | 14.6 | 46.6 | 5.74 | 9.4 | 22.9 | 37.4 |
|  | 200 | 8.2 | 0.09 | 6.9 | 970 | 1603 | 8.1 | 14.6 | 45.3 | 6.99 | 11.2 | 13.4 | 21.4 |
|  | 250 | 14.6 | 0.34 | 8.7 | 870 | 1519 | 9.7 | 15.6 | 43.4 | 4.84 | 7.4 | 4.8 | 7.2 |
|  | 300 | 8.6 | 0.26 | 2.7 | 1000 | 1537 | 2.0 | 2.0 | 2.0 | 1.38 | 2.1 | 1.7 | 2.6 |
|  | 350 | 1.8 | 0.23 | 2.1 | 1200 | 1360 | 2.0 | 2.0 | 2.0 | 1.44 | 2.0 | 1.6 | 2.1 |

$T_{IT}$ (°C): IT temperature, $f\gamma_0$ (vol.%): initial volume fraction of retained austenite, $C\gamma_0$ (mass%): initial carbon concentration of retained austenite, $k$: strain-induced transformation factor, YS (MPa): yield stress or 0.2% offset proof stress, TS (MPa): tensile strength, UEl (%): uniform elongation, TEl (%): total elongation, RA (%): reduction of area, $H_{max}$ (mm): maximum stretch height, HER (%): hole-expansion ratio.

### 3.2. Tensile Properties

Figure 5 shows the typical engineering stress–strain curves and the instantaneous strain-hardening exponent-true strain curves of the 1.5Mn, 3Mn, and 5Mn steels subjected to the IT process at $M_f$ + (30 °C to 70 °C). Figure 6 and Table 2 show the tensile properties of these steels. The 5Mn steel has the highest tensile strength and instantaneous strain-hardening exponent (Figure 5), which result from a high concentration of Mn, the high MA fraction, and the strain-induced transformation of a large amount of retained austenite [30]. In all steels, the diffuse necking occurs at a relatively early strain, namely at half of the total fracture strain (Figure 5a), similar to conventional quenched and tempered martensitic steel. It is noteworthy that severe serration hardly appears on the flow curve of the 5Mn steel, differing from a duplex type 5Mn steel having the same chemical composition [21,36,37].

The 5Mn steel also possesses much larger uniform and total elongations (UEl and TEl) than the 1.5Mn steel at an IT temperature range between 100 °C and 250 °C, as with the 3Mn steel at an IT temperature range between 100 °C and 400 °C (Figure 6b). Note that a difference in UEl between the 1.5Mn and 5Mn steels is smaller than that in the TEl. This indicates that the 5Mn steel is characterized by larger local elongation (LEl = TEl − UEl). When the reduction of area (RA) of these steels was compared in an IT temperature range between $M_s$ and $M_f$, the RA decreases with increasing Mn content (Figure 6c). The 3Mn and 5Mn steels subjected to the DQ process (the IT process at 25 °C) showed considerably low UEl, TEl, LEl, and RA, but the IT process at 100 °C ($<M_f$) yielded high UEl, TEl, LEl, and RA, similar to those subjected to the IT process at the temperatures between $M_s$ and $M_f$.

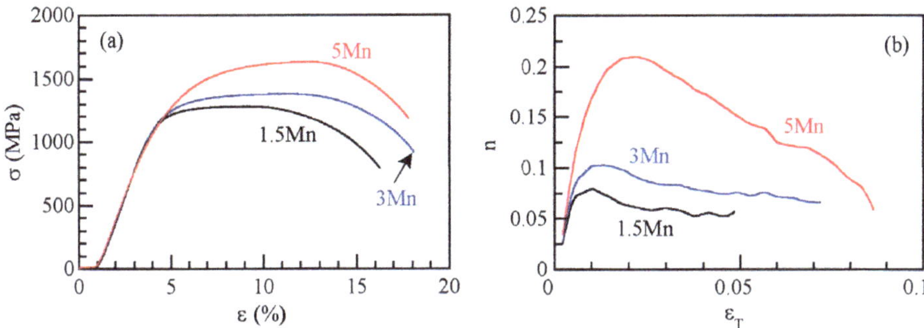

**Figure 5.** (**a**) Typical engineering stress–strain (σ-ε) curves and (**b**) instantaneous strain hardening exponent-true strain (n-ε$_T$) curves of the 1.5Mn, 3Mn, and 5Mn steels subjected to the IT process at $T_{IT}$ = 370 °C, 280 °C, and 180 °C, respectively. The IT temperatures are corresponding to $M_f$ + (30 °C to 70 °C).

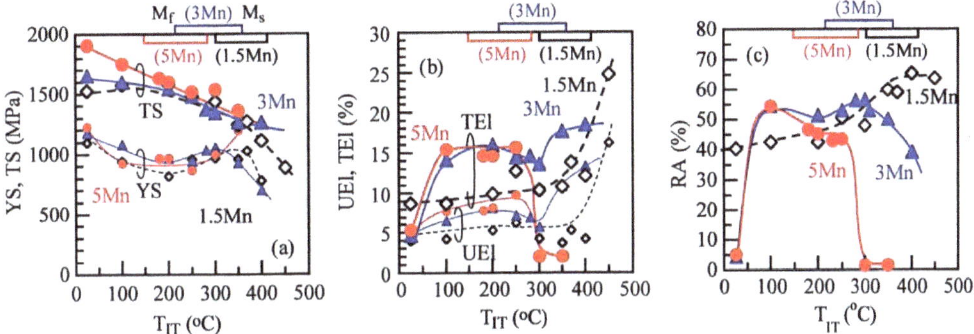

**Figure 6.** Variations in (**a**) yield stress (0.2% offset proof stress) (YS) and tensile strength (TS), (**b**) uniform (UEl) and total elongations (TEl), and (**c**) reduction of area (RA) as a function of IT temperature ($T_{IT}$) in the 1.5Mn (◇), 3Mn (▲), and 5Mn (●) steels.

Figure 7a shows the Mn-content dependences of UEl, TEl, and the products of TS and UEl and TEl (TS × UEl and TS × TEl), which are averaged values at an IT temperature range between $M_s$ and $M_f$ − 100 °C. Both products of the steels considerably increase with increasing Mn content, compared with the UEl and TEl. The TEl of the 5Mn steel is larger than that of the 1.5Mn steel and the same as that of the 3Mn steel, although UEl increases with increasing Mn content.

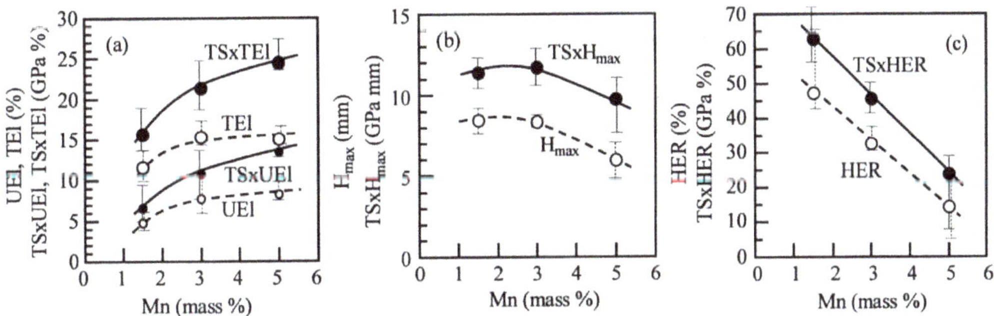

**Figure 7.** Mn content dependences of (**a**) UEl, TS×UEl, TEl, and TS × TEl and (**b**) $H_{max}$ and TS × $H_{max}$, and (**c**) HER and TS × HER which are averaged values of the steels subjected to the IT process at temperatures between $M_s$ and $M_f$ − 100 °C (corresponding to dotted frames in Table 2).

### 3.3. Stretch-Formability

Figure 8a–c shows the typical appearance of the 1.5Mn, 3Mn, and 5Mn steel samples subjected to the IT process at $M_f$ + (30 °C to 70 °C) after stretch-forming tests. The crack initiates near the punch head in all steels. Figure 9a and Table 2 show the $H_{max}$ of these steels. The $H_{max}$ of the 1.5Mn steel increases with increasing IT temperature. The IT temperature dependence of the 3Mn steel shows the same tendency as that of the 1.5Mn steel, except for $T_{IT}$ = 400 °C. The high $H_{max}$s are obtained even in the 1.5Mn and 3Mn steels with a tensile strength above 1500 MPa. The $H_{max}$s of the 5Mn steel are lower than those of the 1.5Mn and 3Mn steels. Moreover, the IT temperature dependence of $H_{max}$ is different from those of the 1.5Mn and 3Mn steels. That is, the $H_{max}$ achieves the peak values at the IT temperatures just above and below $M_f$. When subjected to the DQ process, only the 5Mn steel possesses a much low $H_{max}$.

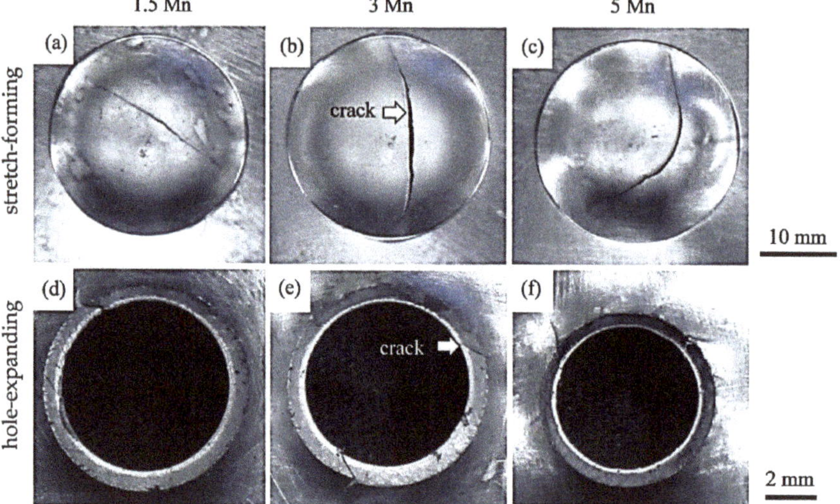


**Figure 8.** Typical samples of the 1.5Mn, 3Mn, and 5Mn steels respectively subjected to the IT process at 370 °C, 280 °C, and 180 °C after stretch-forming (**a**–**c**) and hole-expanding tests (**d**–**f**). The IT temperatures are corresponding to $M_f$ + (30 °C to 70 °C).

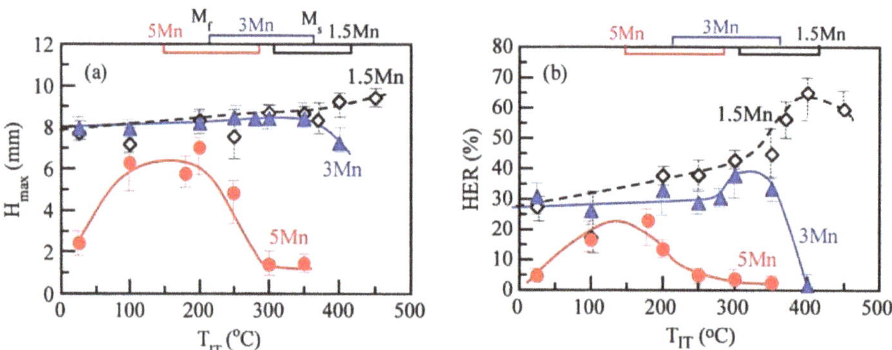


**Figure 9.** Variations in (**a**) maximum stretch height ($H_{max}$) and (**b**) hole-expansion ratio (HER) as a function of isothermal transformation temperature ($T_{IT}$) in the 1.5Mn ($\Diamond$), 3Mn ($\blacktriangle$), and 5Mn ($\bullet$) steels.

Figure 7b shows the average values of the $H_{max}$ and the products of TS and $H_{max}$ (TS × $H_{max}$) of the 1.5Mn, 3Mn, and 5Mn steels subjected to the IT process at the temperatures between $M_s$ and $M_f − 100\,°C$. The Mn content dependence of TS × $H_{max}$ is relatively small, although the peak value is obtained in the 3Mn steel.

### 3.4. Stretch-Flangeability

Figure 10 shows the typical shear stress-displacement curves on hole-punching in the 1.5Mn, 3Mn, and 5Mn steels subjected to the IT process at $M_f + (30\,°C\ to\ 70\,°C)$. The shear stress increases with increasing Mn content, although it decreases after peak shear stress in the 3Mn and 5Mn steels. The final punching displacement of the 1.5Mn steel is slightly larger than those of the 3Mn and 5Mn steels.

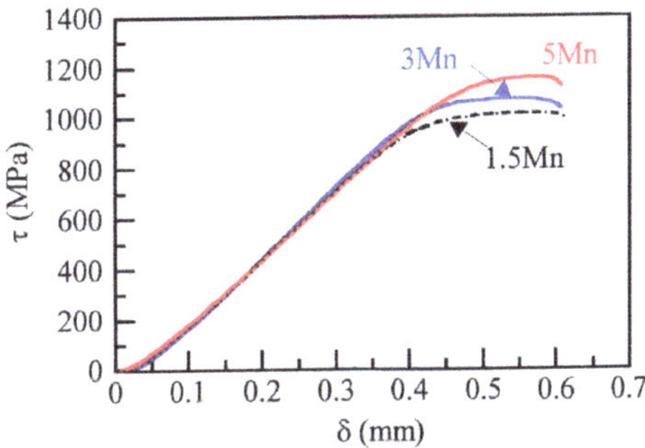


**Figure 10.** Typical shear stress-displacement ($τ$-$δ$) curves on hole-punching in 1.5Mn, 3Mn, and 5Mn steels subjected to the IT process at 370 °C, 280 °C, and 180 °C, respectively. The IT temperatures are corresponding to $M_f + (30\,°C\ to\ 70\,°C)$.

Figure 11 shows SEM images of the break section after hole-punching in the 1.5Mn, 3Mn, and 5Mn steels subjected to the IT process at $M_f + (30\,°C\ to\ 70\,°C)$. A large number of voids are formed in the punching surface layer of these steels, accompanied by significant plastic flow. The shear section length decreases with increasing Mn content (see the bottom of Figure 11). When the void properties are measured in a region from surface to 50 μm depth, the average void diameter increases with increasing Mn content, although the void density decreases with increasing Mn content.

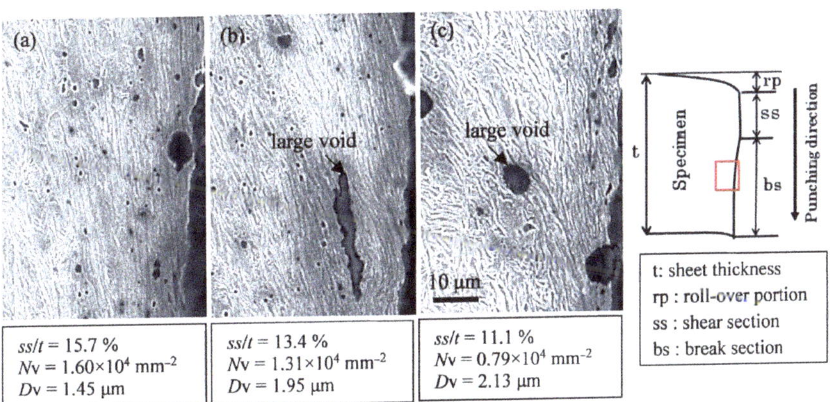


**Figure 11.** SEM images of break section after hole punching of (**a**) the 1.5Mn, (**b**) 3Mn, and (**c**) 5Mn steels subjected to the IT process at $T_{IT} = 370\,°C$, 280 °C, and 180 °C, respectively. The IT temperatures are corresponding to $M_f + (30\,°C\ to\ 70\,°C)$. *ss*/*t*: a ratio of shear section length (*ss*) to thickness (*t*), $N_v$: void density, $D_v$: average void diameter.

Figure 8d–f shows the typical appearance of the 1.5Mn, 3Mn, and 5Mn steel samples subjected to the IT process at $M_f$ + (30 °C to 70 °C) after hole-expanding tests. A few cracks initiate at the punched surface without necking in all steels. Figure 9b and Table 2 show the HER of these steels. The IT temperature dependences of HER of the 1.5Mn, 3Mn, and 5Mn steels resemble those of the $H_{max}$ (Figure 9b). However, these IT temperature dependences appear more prominent than those of the $H_{max}$. In the same way as the UEl, TEl, and RA of Figure 6b,c, the 5Mn steel subjected to the DQ process possesses a much low HER.

Figure 7c shows the average values of the HER and the products of TS and HER (TS × HER) of the 1.5Mn, 3Mn, and 5Mn steels subjected to the IT process at the temperatures between $M_s$ and $M_f$ − 100 °C. The Mn-content dependence of the TS × HER drastically decreases with increasing Mn content, differing from the TS × $H_{max}$. The value of the 5Mn steel is half that of the 1.5Mn steel. This indicates that the 5Mn steel is disfavorable for cold hole expansion despite a good tensile ductility.

## 4. Discussion

In general, stress states of sheet forming can be classified into four modes [38], namely

- deep drawing [stretch (tension) and shrink (compression)],
- stretch-forming [equi-biaxial stretch (tension)],
- stretch-flanging [stretch (tension)] and
- bending [stretch (tension) and bending].

In a case of stretch-flanging, hole-punching by shearing is conducted before hole-expanding. Moreover, these formabilities are influenced by the following microstructural properties in the third generation AHSSs [27–29].

(i)   volume fraction and mechanical stability of retained austenite which control the strain-induced martensite transformation hardening and plastic relaxation of the localized stress concentration (or TRIP effect),

(ii)  MA phase properties such as volume fraction, hardness, etc. which influences the internal stress hardening and void/crack-initiation and -growth behavior at the matrix/MA phase. For the M–MMn steel, the following hardening takes part in the deformation [21],

(iii) Mn concentration or content which contributes to the solid solution hardening and influences the sizes of prior austenitic grain, coarse martensite, and retained austenite, as well as the prior austenitic grain boundary strength.

In the following, the stretch-formability and stretch-flangeability of the 1.5Mn, 3Mn, and 5Mn steels are related to the microstructural properties (i) to (iii), as well as the tensile ductility.

### 4.1. Relationship between Tensile Ductility and Microstructural Properties

The UEl, TS × UEl, TEl, and TS × TEl of the present steels increased with increasing Mn content when they were subjected to the IT process at temperatures between $M_s$ and $M_f$ − 100 °C. The TS × UEl and TS × TEls were constant or decreased with increasing retained austenite fraction (Figure 12a) and increased with increasing $k$-values, except for the TS × TEl of the 5Mn steel (Figure 12b). Such results have been also found in 0.2%C-(1.0−2.5)%Si-(1.0−2.0)%Mn TPF steels tested at room temperature [39]. Sugimoto et al. [39,40] found that the TS × TEl at optimum warm testing temperatures at which the mechanical stability of the retained austenite becomes the maximum linearly increased with increasing initial volume fraction of retained austenite in the TPF steels. So, the results of Figure 12 are considered to be mainly caused by the low mechanical stability of the retained austenite, not the retained austenite fraction.

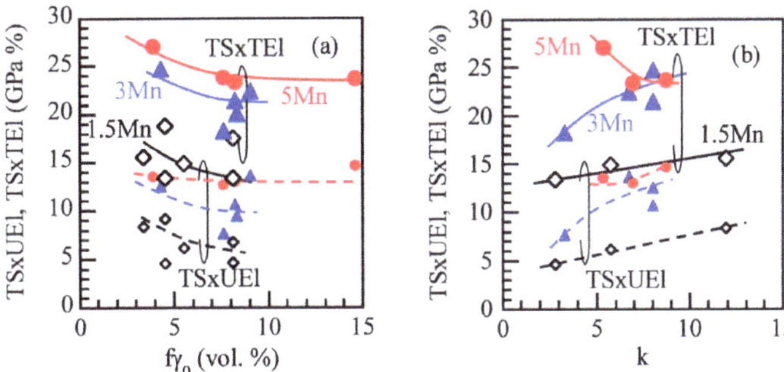

**Figure 12.** Relationships between TS × TEl and TS × UEl and (**a**) initial volume fraction ($f\gamma_0$) and (**b**) strain-induced transformation factor ($k$) of retained austenite in the 1.5Mn (◇), 3Mn (▲), and 5Mn (●) steels. The IT temperatures are between $M_s$ and $M_f - 100\,°C$.

According to Kobayashi et al. [27], Sugimoto et al. [28], and Pham et al. [29] found that a given volume fraction of MA phase fraction (10–15 vol.%) improves the TS × TEl in 0.2%C-1.5%Si-1.5%Mn-1.0%Cr-0.05%Nb TM steel [28]. In the present research, the MA phase fraction increased with increasing Mn content (Figures 3 and 13b). Therefore, the high TS × TEl of the 5Mn steel is considered to be mainly associated with the large strain-hardening behavior (Figure 5b) resulting from the (i), (ii), and (iii), accompanied with the plastic relaxation of the retained austenite which lowers the localized stress concentration at the MA phase/matrix interface and resultantly causes the difficult void/crack-initiation, even high MA phase fraction. The degree of contribution of the (i) to (iii) to the tensile ductility is summarized in Table 3.

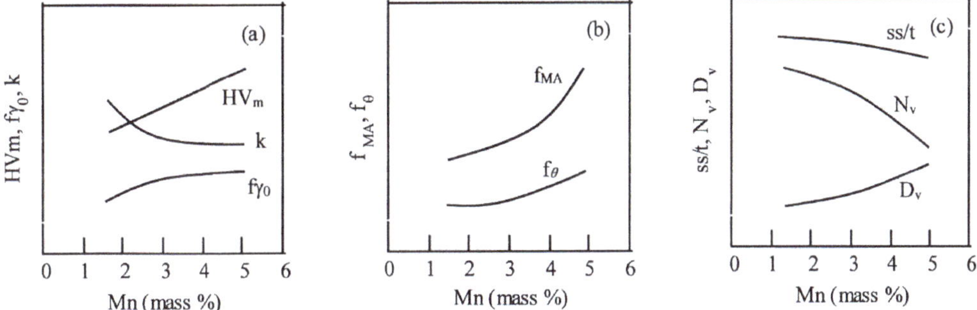

**Figure 13.** Schematic Mn content dependences of (**a**) Vickers hardness of matrix ($HV_m$) and initial volume fraction ($f\gamma_0$) and strain-induced transformation factor ($k$) of retained austenite, (**b**) volume fractions of MA phase ($f_{MA}$) and carbide ($f_\theta$), and (**c**) a ratio of shear section length to sheet thickness ($ss/t$), void density ($N_v$) and average void diameter ($D_v$) in the steels subjected to the IT process at temperatures of $M_f + (30\,°C$ to $70\,°C)$.

**Table 3.** Degree of the contribution of microstructural property to formability in the 5Mn steel.

| Microstructural Property | Tensile Ductility | Stretch-Formability | Stretch-Flangeability | |
|---|---|---|---|---|
| | | | Hole-Punching | Hole-Expanding |
| (i) retained austenite | significant increase | increase | increase | slight increase |
| (ii) MA phase | increase | significant decrease | significant decrease | significant decrease |
| (iii) Mn concentration | increase | increase | unknown | unknown |
| Total of (i) to (iii) | significant increase | decrease | significant decrease | |

In Figure 6b, the UEl and TEl of the 5Mn steel subjected to the DQ process (the IT process at $T_{IT}$ = 25 °C) were significantly small compared to those subjected to the IT process at the temperatures between $M_s$ and $M_f$ − 100 °C. This may be mainly caused by the lower volume fraction and lower carbon concentration (mechanical stability) of retained austenite (Figure 4a,b).

### 4.2. Relationship between Stretch-Formability and Microstructural Properties

As well known, the stretch-formability of the third-generation AHSSs is also controlled by the above (i) to (iii). However, under an equi-biaxial stress state, the strain-induced transformation of the retained austenite and void/crack-initiation at the MA phase/matrix interface are promoted, compared with those under a uniaxial tension-stress state [40]. Resultantly, stretch-formability is significantly influenced by strain-induced transformation and the void/crack-formation behaviors, compared with tensile ductility.

In Figure 7b, the 5Mn steel showed lower $H_{max}$ and TS × $H_{max}$ than the 1.5Mn and 3Mn steels, despite possessing higher tensile ductility. As shown in Figure 14a, the TS × $H_{max}$ of the 1.5Mn and 3Mn steels shows a positive linear relationship with the TS × UEl, although the TS × $H_{max}$ of the 5Mn steel was lower than those of the 1.5Mn and 3Mn steels. As shown in Figure 14b,c, the TS × $H_{max}$ of the present steels decreases with increasing retained austenite fraction and increases with increasing $k$-value, in the same way as the TS × UEl and TS × TEl (Figure 12). The volume fraction and mechanical stability of retained austenite dependences of the TS × $H_{max}$ differ from that reported by Sugimoto et al. [39], while the retained austenite with high mechanical stability considerably increased the $H_{max}$ and TS × $H_{max}$ in 0.2%C-(1.0−2.5)%Si-(1.0−2.0)%Mn TPF steels. Therefore, these dependencies may be caused by the low mechanical stability. In this case, the retained austenite contributes to the increasing of the TS × $H_{max}$. The 5Mn steel contained a larger amount of harder MA phase. Therefore, it is considered that the MA phase promotes the easy void/crack-initiation at the MA phase/matrix interface resulting from the equi-biaxial stress state and mainly decreases the stretch-formability. The solid-solution hardening of Mn maybe also contribute to increasing its stretch-formability. The degree of contribution of the (i) to (iii) on the stretch-formability is summarized in Table 3. It is considered that the carbides in the 5Mn steel (Figure 3f) hardly influence the $H_{max}$ and TS × $H_{max}$ because the amount is small and the crack initiation site is MA phase/matrix interface.

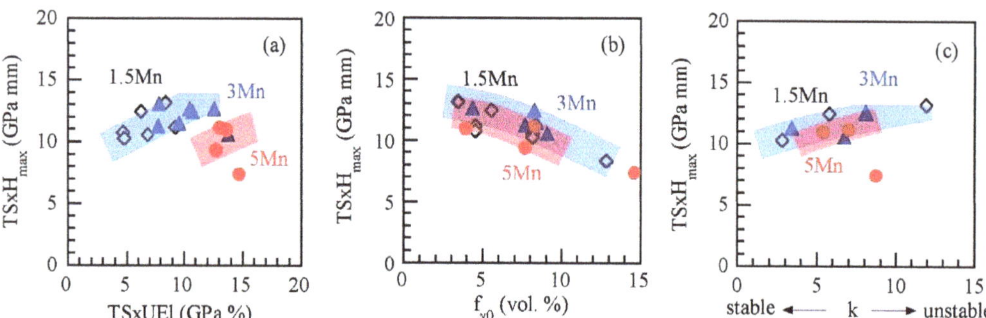


**Figure 14.** Relationships between TS × $H_{max}$ and (**a**) TS × UEl, (**b**) initial retained austenite fraction ($f_{\gamma 0}$), and (**c**) strain-induced transformation factor ($k$) in the 1.5Mn (◇), 3Mn (▲), and 5Mn (•) steels. IT temperatures are between $M_s$ and $M_f$ − 100 °C.

### 4.3. Relationship between Stretch-Flangeability and Microstructural Properties

According to Sugimoto et al. [7,8,26−28], the stretch-flangeability of the third generation AHSS sample can be mainly evaluated by

(1)  surface layer damage suffered on hole-punching (severe plastic flow, strain hardening behavior, and void/crack-initiation damage) and

(2)  ductility and crack/void-growth and -connection behaviors on hole-expanding.

The (1) and (2) are controlled by the various metallurgical characteristics, such as hardness of the matrix structure, MA phase properties, a hardness ratio of the second phase to the matrix, micro-scale uniformity, retained austenite characteristics, carbide properties, etc. In the third-generation AHSS, the (1) plays a dominating role in the stretch-flangeability. When the 1.5Mn, 3Mn, and 5Mn steels were subjected to the IT process at the temperatures between $M_s$ and $M_f - 100\ °C$, the averaged values of HER and TS × HER linearly decreased with increasing Mn content (Figure 7c). In the following, the relationship between the stretch-flangeability and the metallurgical characteristics is discussed.

First, let us discuss the effect of hole-punching damage on stretch-flangeability. When the present steels were subjected to the IT process at the temperatures between $M_s$ and $M_f - 100\ °C$, the 3Mn and 5Mn steels possessed a larger amount of stable retained austenite than the 1.5Mn steel (Figures 4 and 13a). Also, the microstructure of the 5Mn steel can be characterized by higher matrix hardness (lower image quality in Figure 3c) and higher volume fractions of MA phase and carbide than those of the 1.5Mn steel (Figures 3 and 13b). The hardness ratio of the MA phase to the coarse martensite matrix structure is estimated to be relatively high because the image quality of the MA phase is significantly low (Figure 3c). Such microstructural characteristics may promote the void/crack-initiation on hole-punching. The largest voids were formed in a hole-punched surface layer of the 5Mn steel, although the number of voids was minimum (Figures 11 and 13c). Moreover, the shear section length was the smallest. Thus, the high MA phase fraction and the high hardness ratio may result in short shear-section length and large void sizes on hole-punching in the 5Mn steel. As shown in Figure 10, the final punching displacement of the 5Mn steel was slightly smaller than that of 1.5Mn steel, although the punching shear stress was higher than that of 1.5Mn steel. This means that the TRIP effect of the retained austenite contributes to the reduction of the punching damage (see Table 3).

Next, we discuss the effect of hole-expanding on the stretch-flangeability of the 5Mn steel. In general, hole-expanding of the third generation AHSSs is mainly controlled by void/crack growth and connection [27–29]. As most retained austenites near the punched-hole surface transform to martensite on hole-punching, the untransformed retained austenite is difficult to contribute to the suppression of the void/crack-growth and -connection on hole-expanding. Therefore, it is considered that the hole-surface damage on punching resulting from the MA phase (referring to the (ii)) may control its stretch-flangeability, with a slight contribution from the retained austenite (see Table 3).

The TS × HER increased with increasing TS × RA in the present steels (Figure 15a). In addition, the TS × HER increases with increasing initial volume fraction of retained austenite (the line I in Figure 15b) and decreasing $k$-value (or increasing mechanical stability of the retained austenite (Figure 15c). This indicates that the retained austenite plays a positive role in its stretch-flangeability by suppressing the void-initiation on hole-punching, because it plays a small role in hole-expanding. Therefore, the (ii) preferentially decreases the HER by the large hole-punching damage and easy void-growth and -connection on hole-expanding in the 5Mn steel (Table 3), despite the positive contribution from a large amount of retained austenite. As carbide fraction was relatively small, the effect of carbide on stretch-flangeability is considered to be negligible. Unfortunately, the effect of the (iii) on stretch-flangeability is not clear at present.

At the present stage, the discussion about the relationship between stretch-flangeability and the microstructural properties is not perfect, in the same way as the stretch-formability. Further detailed investigation of the relationship is expected in the future.

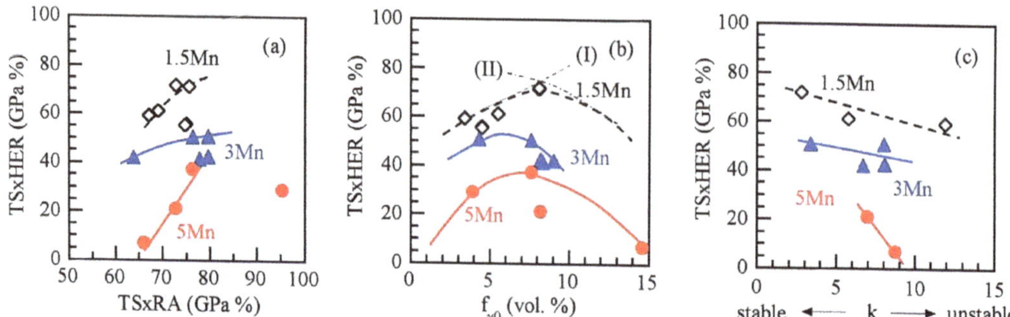

**Figure 15.** Relationships between TS × HER and (**a**) TS × RA, (**b**) initial retained austenite fraction ($f\gamma_0$), and (**c**) strain-induced transformation factor ($k$) in the 1.5Mn ($\diamond$), 3Mn ($\blacktriangle$), and 5Mn ($\bullet$) steels. The IT temperatures are between $M_s$ and $M_f - 100\,°C$. Dashed lines I and II in (**b**) represent positive and negative $f\gamma_0$ dependences of TS × HER, respectively.

### 4.4. Further Improvement of Stretch-Formability and Stretch-Flangaebility

Finally, we discuss how to improve the $H_{max}$ and HER of the 5Mn steel. According to Sugimoto et al. [35], warm hole-punching and/or hole-expanding at 200 °C significantly increase the TS × HER of duplex type 3Mn and 5Mn steels with reverted austenites of 24 and 40 vol.%, respectively. Although the present martensite-type 3Mn and 5Mn steels possess relatively low retained-austenite fractions, it is expected that warm hole-punching and/or hole-expanding improve the stretch-flangeability because the warm working or heating increases the mechanical stability of the retained austenite and softens the matrix and MA phase. Resultantly, the void/crack-initiation, -growth, and -connection are suppressed and shear-section length is increased. In the same way, warm forming is expected to increase the stretch-formability.

In this work, the DQ process considerably decreased the $H_{max}$ and HER of the 5Mn steel. According to Kobayashi et al. [27], partitioning (or tempering) after the DQ process promotes carbon enrichment in the retained austenite and softening of the coarse and fine martensites, with a slight increase in carbide fraction. Thus, this partitioning is also considered to enhance the $H_{max}$ and HER of the 5Mn steel subjected to the DQ process. According to Zheng et al. [41], the easiest way to improve the $H_{max}$ and HER of the medium Mn steel is lowering the carbon content.

### 5. Conclusions

The cold stretch-formability and stretch-flangeability of the martensite type 5Mn steel subjected to the IT process at the temperatures from above $M_s$ to below $M_f$ were investigated and were related to the microstructural properties, as well as tensile ductility. The main results can be summarized as follows:

(1) The highest UEl and TEl were achieved in the 5Mn steel subjected to the IT process at temperatures between $M_s$ and $M_f - 100\,°C$. The TS × UEl and TS × TEl were two and one-half times those of the 1.5Mn steel (equivalent to the TM steel), respectively. This was mainly associated with the TRIP effect of a large amount of retained austenite, accompanied by high strain-hardening by a large amount of MA phase and high Mn content.

(2) High $H_{max}$ was obtained in the 5Mn steel subjected to the IT process at temperatures between $M_s$ and $M_f - 100\,°C$. However, the TS × $H_{max}$ was slightly lower than those of the 1.5Mn and 3Mn steels. In this case, a large amount of metastable austenite and high Mn content contributed to increasing stretch-formability. However, the presence of a large amount of MA phase significantly reduced stretch-formability through facilitating void/crack initiation and growth.

(3) In the 5Mn steel, the IT temperature dependence of the HER resembled that of the $H_{max}$. However, the optimum TS × HER obtained by the IT process at the

temperatures between $M_s$ and $M_f - 100\,^\circ\text{C}$ considerably decreased compared to those of the 1.5Mn and 3Mn steels. The decreased stretch-flangeability of the 5Mn steel was mainly associated with large hole-punching surface damage, such as short shear length and the presence of large voids, mainly caused by the high volume fraction and hardness of the MA phase. The retained austenite contributed to lowering the hole-punching damage. The effect of Mn content on the stretch-flangeability was not clear at present.

**Author Contributions:** The first draft of the paper was written by K.-i.S. and the final editing was done by K.-i.S., H.T. and J.K. All authors have read and agreed to the published version of the manuscript.

**Funding:** This research received no external funding.

**Institutional Review Board Statement:** Not applicable.

**Informed Consent Statement:** Not applicable.

**Data Availability Statement:** Not applicable.

**Acknowledgments:** We thank Tomohiko Hojo from Tohoku University for the kind discussion.

**Conflicts of Interest:** The authors declare no conflict of interest.

## References

1. Rana, R.; Singh, S.B. *Automotive Steels—Design, Metallurgy, Processing and Applications*; Woodhead Publishing: Cambridge, UK, 2016.
2. Fan, D.; Fonstein, N.; Jun, H. Effect of microstructure on tensile properties and cut-edge formability of DP, TRIP, Q&T and Q&P steels. *AIST Trans.* **2016**, *13*, 180–184.
3. Frómeta, D.; Parareda, S.; Lara, A.; Grifé, L.; Tarhouni, I.; Casellas, D. A new cracking resistance index based on fracture mechanics for high strength sheet metal ranking. *IOP Conf. Ser. Mater. Sci. Eng.* **2021**, *1157*, 012094. [CrossRef]
4. Sugimoto, K. Recent progress of low and medium carbon advanced martensitic steels. *Metals* **2021**, *11*, 652. [CrossRef]
5. Zackay, V.F.; Parker, E.R.; Fahr, D.; Bush, R. The enhancement of ductility in high-strength steels. *Trans. Am. Soc. Met.* **1967**, *60*, 252–259.
6. Grässel, O.; Krüger, L.; Frommeyer, G.; Meyer, L.W. High strength Fe-Mn-(Al, Si) TRIP/TWIP steels development—Properties—Application. *Int. J. Plast.* **2000**, *16*, 1391–1409. [CrossRef]
7. Sugimoto, K.; Kanda, A.; Kikuchi, R.; Hashimoto, S.; Kashima, T.; Ikeda, S. Ductility and formability of newly developed high strength low alloy TRIP-aided sheet steels with annealed martensite matrix. *ISIJ Int.* **2002**, *42*, 910–915. [CrossRef]
8. Sugimoto, K.; Sakaguchi, J.; Iida, T.; Kashima, T. Stretch-flangeability of a high-strength TRIP type bainitic steel. *ISIJ Int.* **2000**, *40*, 920–926. [CrossRef]
9. Sugimoto, K.; Tsunezawa, M.; Hojo, T.; Ikeda, S. Ductility of 0.1-0.6C-1.5Si-1.5Mn ultra high-strength low-alloy TRIP-aided sheet steels with bainitic ferrite matrix. *ISIJ Int.* **2004**, *44*, 1608–1614. [CrossRef]
10. Sugimoto, K.; Murata, M.; Song, S. Formability of Al-Nb bearing ultrahigh-strength TRIP-aided sheet steels with bainitic ferrite and/or martensite matrix. *ISIJ Int.* **2010**, *50*, 162–168. [CrossRef]
11. Speer, J.G.; De Moor, E.; Findley, K.O.; Matlock, B.C.; De Cooman, B.C.; Edmonds, D.V. Analysis of microstructure evolution in quenching and partitioning automotive sheet steel. *Metall. Mater. Trans. A* **2011**, *42A*, 3591–3601. [CrossRef]
12. Wang, L.; Speer, J.G. Quenching and partitioning steel heat treatment. *Metallogr. Microstruc. Anal.* **2013**, *2*, 268–281. [CrossRef]
13. Im, Y.; Kim, E.; Song, T.; Lee, J.; Suh, D. Tensile properties and stretch-flangeability of TRIP steels produced by quenching and partitioning (Q&P) process with different fractions of constituent phases. *ISIJ Int.* **2021**, *61*, 572–581.
14. Garcia-Mateo, C.; Paul, G.; Somani, M.C.; Porter, D.A.; Bracke, L.; Latz, A.; De Andres, C.G.; Caballero, F.G. Transferring nanoscale bainite concept to lower contents: A prospective. *Metals* **2017**, *7*, 159. [CrossRef]
15. Tian, J.; Xu, G.; Zhou, M.; Hu, H. Refined bainite microstructure and mechanical properties of a high-strength low-carbon bainitic steel. *Steel Res. Int.* **2018**, *89*, 1700469. [CrossRef]
16. Rana, R.; Halder, S.; Das, S. Mechanical properties of a bainitic steel producible by hot rolling. *Arch. Metall. Mater.* **2017**, *62*, 2331–2338. [CrossRef]
17. Miller, R.I. Ultrafine-grained microstructures and mechanical properties of alloy steels. *Metall. Trans.* **1972**, *3*, 905–912. [CrossRef]
18. Furukawa, T. Dependence of strength-ductility characteristics on thermal history in low carbon, 5 wt-%Mn steels. *Mater. Sci. Technol.* **1989**, *5*, 465–470. [CrossRef]
19. Cao, W.; Wang, C.; Shi, J.; Wang, M.; Hui, W.; Dong, D. Microstructure and mechanical properties of Fe–0.2C–5Mn steel processed by ART-annealing. *Mater. Sci. Eng. A* **2011**, *528*, 6661–6666. [CrossRef]

20. Seo, E.; Cho, L.; De Cooman, B.C. Application of quenching and partitioning processing to medium Mn Steel. *Metall. Mater. Trans. A* **2015**, *46A*, 27–31. [CrossRef]
21. Tanino, H.; Horita, M.; Sugimoto, K. Impact toughness of 0.2 pct-1.5 pct-(1.5 to 5) pct Mn transformation-induced plasticity-aided steels with an annealed martensite matrix. *Metall. Mater. Trans. A* **2016**, *47A*, 2073–2080. [CrossRef]
22. Bleck, W.; Guo, X.; Ma, Y. The TRIP effect and its application in cold formable sheet steels. *Steel Res. Int.* **2017**, *88*, 1700218. [CrossRef]
23. Cao, W.; Zhang, M.; Huang, C.; Xiao, S.; Dong, H.; Weng, Y. Ultrahigh Charpy impact toughness (−450J) achieved in high strength ferrite/martensite laminated steels. *Sci. Rep.* **2017**, *7*, 41459. [CrossRef] [PubMed]
24. Xu, Y.; Hu, Z.; Zou, Y.; Tan, X.; Han, D.; Chen, S.; Ma, D.; Misra, R.D.K. Effect of two-step intercritical annealing on microstructure and mechanical properties of hot-rolled medium manganese TRIP steel containing δ-ferrite. *Mater. Sci. Eng. A* **2017**, *688*, 40–55. [CrossRef]
25. Liu, L.; Yu, Q.; Wang, Z.; Ell, J.; Huang, M.X.; Ritchie, R.O. Making ultrastrong steel tough by grain-boundary delamination. *Science* **2020**, *368*, 1347–1352. [CrossRef]
26. Kim, J.; Kwon, M.; Lee, J.; Lee, S.; Lee, K.; Suh, D. Influence of isothermal treatment prior initial quenching of Q&P process on microstructure and mechanical properties of medium Mn steel. *ISIJ Int.* **2021**, *61*, 518–526.
27. Kobayashi, J.; Pham, D.V.; Sugimoto, K. Stretch-flangeability of 1.5 GPa grade TRIP-aided martensitic cold rolled sheet steels. In Proceedings of the 10th International Conference on Technology of Plasticity (ICTP 2011), Aachen, Germany, 25–30 September 2011; pp. 598–603.
28. Sugimoto, K.; Kobayashi, J.; Pham, D.V. Advanced ultrahigh-strength TRIP-aided martensitic sheet steels for automotive applications. In Proceedings of the New Developments in Advanced High Strength Sheet Steels (AIST 2013), Vail, CO, USA, 23–27 June 2013; pp. 175–184.
29. Pham, D.V.; Kobayashi, J.; Sugimoto, K. Effects of microalloying on stretch-flangeability of TRIP-aided martensitic sheet steel. *ISIJ Int.* **2014**, *54*, 1943–1951. [CrossRef]
30. Hanamura, T.; Torizuka, S.; Tamura, S.; Enokida, S.; Takechi, H. Effect of austenite grain size on transformation behavior microstructure and mechanical properties of 0.1C-5Mn martensitic steel. *ISIJ Int.* **2013**, *53*, 2218–2225. [CrossRef]
31. Sugimoto, K.; Tanino, H.; Kobayashi, J. Impact toughness of medium-Mn transformation-induced plasticity-aided steels. *Steel Res. Int.* **2015**, *86*, 1151–1160. [CrossRef]
32. He, B.; Liu, L.; Huang, M.X. Room-temperature quenching and partitioning steel. *Metall. Mater. Eng. A* **2018**, *349A*, 3167–3172. [CrossRef]
33. He, B.; Wang, M.; Huang, M.X. Improving tensile properties of room-temperature quenching and partitioning steel by dislocation engineering. *Metall. Mater. Trans. A* **2019**, *50A*, 4021–4026. [CrossRef]
34. Maruyama, H. X-ray measurement of retained austenite. *Jpn. Soc. Heat Treat.* **1977**, *17*, 198–204.
35. Dyson, D.J.; Holmes, B. Effect of alloying additions on the lattice parameter of austenite. *J. Iron Steel Inst.* **1970**, *208*, 469–474.
36. Sugimoto, K.; Hidaka, S.; Tanino, H.; Kobayashi, J. Effects of Mn content on the warm stretch-flangeability of C-Si-Mn TRIP-aided steels. *Steel Res. Int.* **2017**, *83*, 1600482. [CrossRef]
37. Sugimoto, K.; Hidaka, S.; Tanino, H.; Kobayashi, J. Warm formability of 0.2 pct C-1.5 pct Si-5 pct Mn transformation-induced plasticity-aided steel. *Metall. Mater. Trans. A* **2017**, *48A*, 2237–2246. [CrossRef]
38. Takahashi, M. Development of high strength steels for automobiles. *Nippon Steel Tech. Rep.* **2003**, *88*, 2–7.
39. Sugimoto, K.; Usui, N.; Kobayashi, M.; Hashimoto, S. Effects of volume fraction and stability of retained austenite on ductility of TRIP-aided dual-phase steels. *ISIJ Int.* **1992**, *32*, 1311–1318. [CrossRef]
40. Sugimoto, K.; Kobayashi, M.; Nagasaka, A.; Hashimoto, S. Warm stretch-formability of TRIP-aided dual-phase sheet steels. *ISIJ Int.* **1995**, *35*, 1407–1414. [CrossRef]
41. Zheng, G.; Chang, Y.; Fan, Z.; Li, X.; Wang, C. Study of thermal forming limit of medium-Mn steel based on finite element analysis and experiments. *Int. J. Manuf. Technol.* **2018**, *94*, 133–144. [CrossRef]

*Article*

# Influence of Cooling Process Routes after Intercritical Annealing on Impact Toughness of Duplex Type Medium Mn Steel

**Koh-ichi Sugimoto * and Hikaru Tanino**

School of Science and Technology, Shinshu University, 4-17-1 Wakasato, Nagano 380-8553, Japan; 14tm116j@shinshu-u.ac.jp
* Correspondence: sugimot@shinshu-u.ac.jp; Tel.: +81-90-9667-4482

**Abstract:** To apply the duplex type low-carbon medium-manganese steel to the hot/warm-forging and -stamping products, the influence of cooling process routes immediately after intercritical annealing such as air-cooling (AC) and isothermal transformation (IT) processes on the impact toughness of 0.2%C-1.5%Si-5%Mn (in mass %) duplex type medium-Mn (D-MMn) steel was investigated. Moreover the microstructural and tensile properties were also investigated. The AC process increased the volume fraction of reverted austenite but decreased the thermal and mechanical stability in the D-MMn steel, compared to the IT process. The AC process increased the tensile strength but decreased the total elongation. The Charpy V-notch impact value and ductile-brittle transition temperature were deteriorated by the AC process, compared to the IT process. This deterioration of the impact toughness was mainly related to the reverted austenite characteristics and fracture mode.

**Keywords:** heat treatment; microstructure; reverted austenite; tensile property; impact toughness; duplex type medium Mn steel

**Citation:** Sugimoto, K.-i.; Tanino, H. Influence of Cooling Process Routes after Intercritical Annealing on Impact Toughness of Duplex Type Medium Mn Steel. *Metals* **2021**, *11*, 1143. https://doi.org/10.3390/met11071143

Received: 15 June 2021
Accepted: 18 July 2021
Published: 20 July 2021

**Publisher's Note:** MDPI stays neutral with regard to jurisdictional claims in published maps and institutional affiliations.

## 1. Introduction

The first-, second-, and third-generation advanced high strength steels (AHSSs) recently developed are expected to apply to the automotive sheet products [1–3] and wire-rod products [4–7]. Among them, the third-generation AHSSs such as Si/Al-Mn transformation-induced plasticity (TRIP)-aided bainitic ferrite (TBF) [8,9] and martensite (TM) [10,11] steels, carbide-free bainitic (CFB) steel [12,13], and quenching and partitioning (Q&P) steel [14,15] bring on great weight reduction and high reliability due to the ultrahigh strength above 980 MPa. The microstructure of the third-generation AHSSs mainly consists of the matrix structure of bainitic ferrite/martensite containing metastable retained austenite of 5–20 vol. %. To obtain a larger amount of reverted austenite, annealed martensite/reverted austenite duplex type medium Mn (D-MMn) steels with 4–14 mass % Mn [16–21] were also developed as well as martensite-type medium Mn (M-MMn) steels [22–25] and laminate type medium Mn (L-MMn) steels [25,26].

The D-MMn steel can easily achieve a higher product of tensile strength and total elongation (TS × TEl) than 30 GPa% through intercritical annealing and then air cooling (AC) or quenching in a water/oil bath [16–21]. The D-MMn steel subjected to the AC process possesses a high-impact absorbed value [27–30]. Thus, some applications of the D-MMn steel to forging and stamping products [31,32] are reported up to now. Tanino et al. [21] proposed that an isothermal transformation (IT) process immediately after intercritical annealing enhances further the TS × TEl and impact absorbed value through the TRIP effect or strain-induced martensite transformation (SIMT) hardening of a large amount of metastable reverted austenite in 0.2%C-1.5%Si-5%Mn (in mass %) D-MMn steel. However, the tensile properties and impact toughness of the D-MMn steels produced with the AC process are not exactly compared with those subjected to the IT process.

In the present study, the microstructure, tensile properties, and impact toughness of the 0.2%C-1.5%Si-5%Mn D-MMn steel subjected to the AC process after intercritical annealing were compared with those of the steel subjected to the IT process for applications to hot/warm-forging and stamping products. In addition, the relation between the impact toughness and microstructural properties, especially metastable reverted austenite characteristics, was discussed.

## 2. Materials and Methods

In this study, medium Mn steel with Mn content of 5 mass % (5 Mn steel) was prepared as 100 kg slabs via vacuum melting. For comparison, slabs containing Mn contents of 1.5 and 3 mass % (1.5 Mn and 3 Mn steels, respectively) were also produced in the same conditions. The chemical compositions of these slabs are shown in Table 1. These slabs were heated to 1200 °C and then hot-rolled to a thickness of 5 mm with a finishing temperature of 850 °C. The hot-rolled plates were then ground to a thickness of 3 mm. The austenitic and ferritic transformation temperatures ($Ac_3$ and $Ac_1$, respectively) and the martensite start and finish temperatures ($M_s$ and $M_f$, respectively) of these steels were determined using a dilatometer (Thermecmaster-Z, Fuji Electronic Ind. Co., Osaka, Japan). The measured continuous cooling transformation (CCT) curves of the steels are shown in Figure 1.

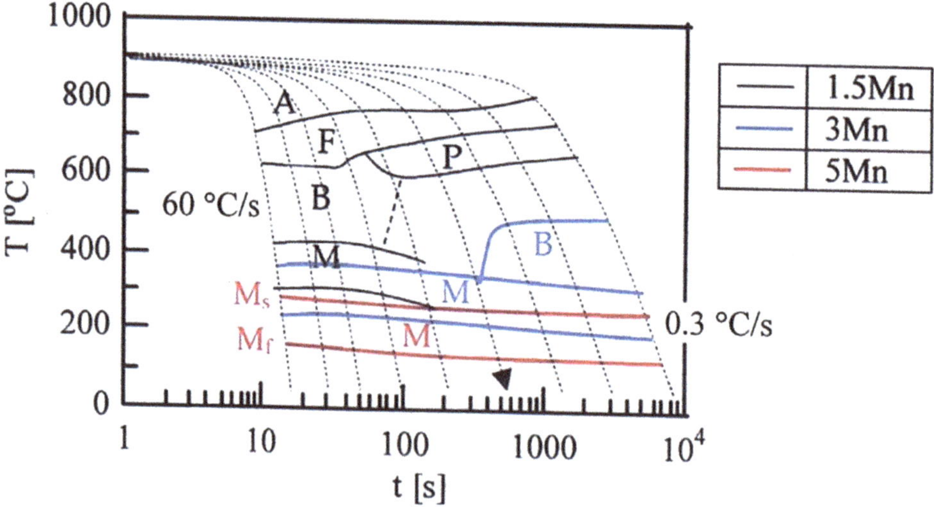


**Figure 1.** CCT diagrams of the 1.5 Mn, 3 Mn and 5 Mn steels. "A", "F", "P", "B" are austenite, ferrite, pearlite, bainite, and martensite, respectively.

**Table 1.** Chemical composition (mass %) and austenite-start and -finish temperatures and martensite-start and -finish temperatures ($Ac_1$, $Ac_3$, $M_s$, and $M_f$ in °C) of steels used.

| Steel | C | Si | Mn | P | S | Al | N | O | $Ac_1$ | $Ac_3$ | $M_s$ | $M_f$ |
|---|---|---|---|---|---|---|---|---|---|---|---|---|
| 1.5 Mn | 0.20 | 1.49 | 1.50 | 0.006 | 0.0015 | 0.035 | 0.0038 | <0.001 | 719 | 847 | 420 | 300 |
| 3 Mn | 0.20 | 1.52 | 2.98 | 0.006 | 0.0016 | 0.037 | 0.0034 | <0.001 | 689 | 797 | 363 | 220 |
| 5 Mn | 0.21 | 1.50 | 4.94 | 0.005 | 0.0016 | 0.032 | 0.0020 | <0.001 | 657 | 741 | 282 | 150 |

Specimens for tensile tests (JIS-14B, 2.5-mm thick, 25 mm gauge in length, 4 mm wide) and sub-sized V-notched impact tests (JIS-5, 55 mm long, 10 mm wide, 2.5 mm thick) were machined from the plates along the rolling direction. The specimens were subsequently subjected to the heat-treatment process shown in Figure 2; (1) austenitizing at 800 to 900 °C and then quenching in an oil bath, (2) intercritical annealing between $Ac_1$ and $Ac_3$ for 1200 s at 780, 720, and 680 °C for the 1.5 Mn, 3 Mn, and 5 Mn sheets of steel, respectively, and (3) the AC process and IT process at $T_{IT} = M_s - 100\ °C$ for IT times, $t_{IT}$, of 500, 1000, and 5000 s in salt and an oil bath. The intercritical annealing temperatures were chosen as the temperatures at which the ferrite and austenite volume fractions are 50% and 50%, respectively. The IT temperatures and IT times that give optimum reverted austenite characteristics were selected from previous research [21].

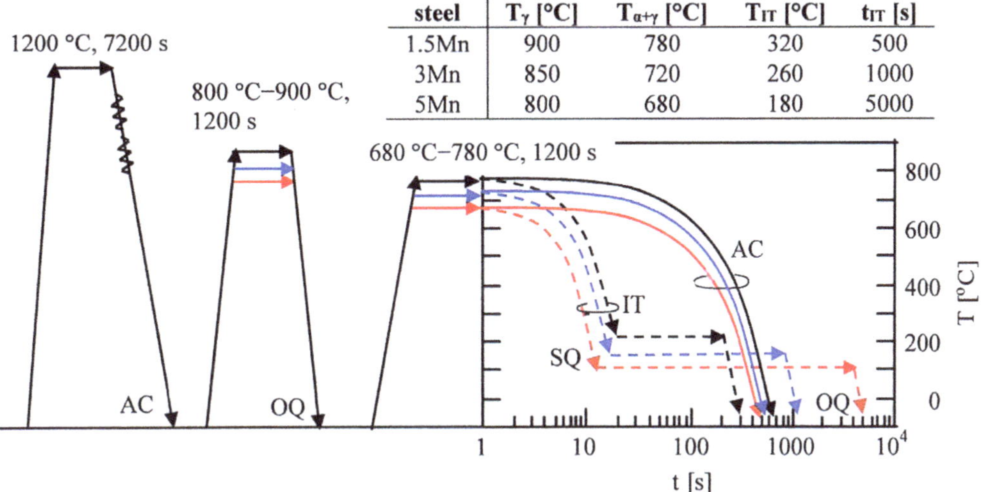



| steel | $T_\gamma$ [°C] | $T_{\alpha+\gamma}$ [°C] | $T_{IT}$ [°C] | $t_{IT}$ [s] |
|---|---|---|---|---|
| 1.5Mn | 900 | 780 | 320 | 500 |
| 3Mn | 850 | 720 | 260 | 1000 |
| 5Mn | 800 | 680 | 180 | 5000 |

**Figure 2.** Heat treatment diagram of the AC and IT processes in the 1.5 Mn (black line), 3 Mn (blue line), and 5 Mn (red line) steels. OQ: quenching in an oil bath (53.5 °C/s), SQ: quenching in a salt bath (24.0 °C/s), AC: air cooling (1.2 °C/s).

The microstructure of the steels was observed by field-emission scanning electron microscopy (FE-SEM; JSM-7000F, JEOL Ltd., Akishima, Tokyo, Japan) which was performed using an electron backscatter diffraction system (EBSD; OIM system, TexSEM Laboratories, Inc., Prova, UT, USA) with the step size of 0.1 μm. The steel specimens for the FE-SEM–EBSD analyses were first ground with alumina powder and colloidal silica and then prepared by ion thinning.

The characteristics of the reverted austenite in the steel samples were quantified by X-ray diffraction (XRD; RINT2100, Rigaku Co., Akishima, Tokyo, Japan). The specimens were electropolished after being ground with emery paper (#2000). The volume fraction of the reverted austenite ($f\gamma$, vol. %) was calculated from the integration of the intensities of the (200)α, (211)α, (200)γ, (220)γ, and (311)γ peaks of the XRD patterns obtained using the Cu Kα radiation [33]. The carbon concentration ($C\gamma$, mass %) was estimated by substituting the lattice constant ($a_\gamma$; unit of $10^{-1}$ nm) measured from the (200)γ, (220)γ, and (311)γ peaks of the Cu-Kα radiation into the empirical formula proposed by Dyson and Holmes [34]. For convenience, the contents of the added alloying elements were substituted for these concentrations in this study. In this research, the average values of volume fractions and carbon concentrations of reverted austenite measured at three locations were adopted, as well as the lath size of annealed martensite structure.

The thermal stability of the reverted austenite was evaluated in a temperature range between −196 °C and 300 °C by measuring the volume fraction change of the reverted austenite transformed on cooling in dry ice, ethyl alcohol, and/or liquid nitrogen, and on heating in water and using a pair of plate heaters (70 × 90 mm²). A holding time of 1200 s was used for cooling and heating. The mechanical stability of the reverted austenite was defined using "the strain-induced transformation factor $k$" in the following equation [11,21,35],

$$\ln f\gamma = \ln f\gamma_0 - k\,\varepsilon \tag{1}$$

where $f\gamma_0$ and $f\gamma$ are the volume fractions of reverted austenite before and after straining to the true plastic strain ($\varepsilon$) via tension, respectively.

Tensile tests were carried out at 25 °C using a tensile testing machine (AG-10TD, Shimadzu Co., Kyoto) under a crosshead speed of 1 mm/min (resulting in a strain rate of $6.67 \times 10^{-4}\,\mathrm{s}^{-1}$). The impact tests were conducted on conventional and instrumented Charpy impact testing machines (CI-300 and CAI300, Tokyo Testing Machine Inc., Tokyo, Japan) for temperatures in the range of −196 °C to 100 °C. Liquid nitrogen, dry ice, ethyl alcohol, and water were used to cool and heat the specimens. The specimens were held at different testing temperatures for 1800 s before being tested. The impact tests were performed within 3 s after removing the specimen from the temperature-regulating medium. The impact properties were evaluated by determining the Charpy V-notch impact value ($E_v$) and 50% shear fracture ductile-to-brittle transition temperature (DBTT) of the specimens. At least three tensile and impact specimens were tested for each condition to obtain the average values of the tensile properties and $E_v$ at 25 °C.

## 3. Results

### 3.1. Microstructure and Reverted Austenite Characteristics

Figure 3 shows the microstructures of the 1.5 Mn, 3 Mn, and 5 Mn steels subjected to the AC and IT processes. The microstructures of the steels subjected to the AC process are nearly the same as those of the steels subjected to the IT process. The microstructures mainly consist of an annealed martensite structure matrix and metastable reverted austenite. The lath size of the annealed martensite structure measured by the line intersecting method is nearly the same in the cases of the AC and IT processes, although it considerably decreases with increasing Mn content. Martensite–austenite (MA) phase exists only in the 1.5 Mn steel. The size of the reverted austenite is nearly the same in all steels, although the volume fraction increases with increasing Mn content. According to previous research [21], the dislocation density of the annealed martensite structure of the 5 Mn steel is very low in the same way as those of the 1.5 Mn and 3 Mn steels.

Figure 4 and Table 2 show the reverted austenite characteristics of the steels subjected to the AC and IT processes. The AC process decreases the carbon concentration and mechanical stability of the reverted austenite compared to the IT process although it increases the volume fraction of the reverted austenite. The low carbon concentration decreased by the AC process is considered to be associated with the short carbon-enrichment time of the reverted austenite. The reverted austenite fraction increases with increasing Mn content in both processes. On the other hand, the carbon concentration decreases with increasing Mn content. Unlike the Mn content dependence of carbon concentration, the 3 Mn steel shows the highest $k$-value and the 5 Mn steel exhibits an intermediate $k$-value between the 1.5 Mn and 3 Mn steels.

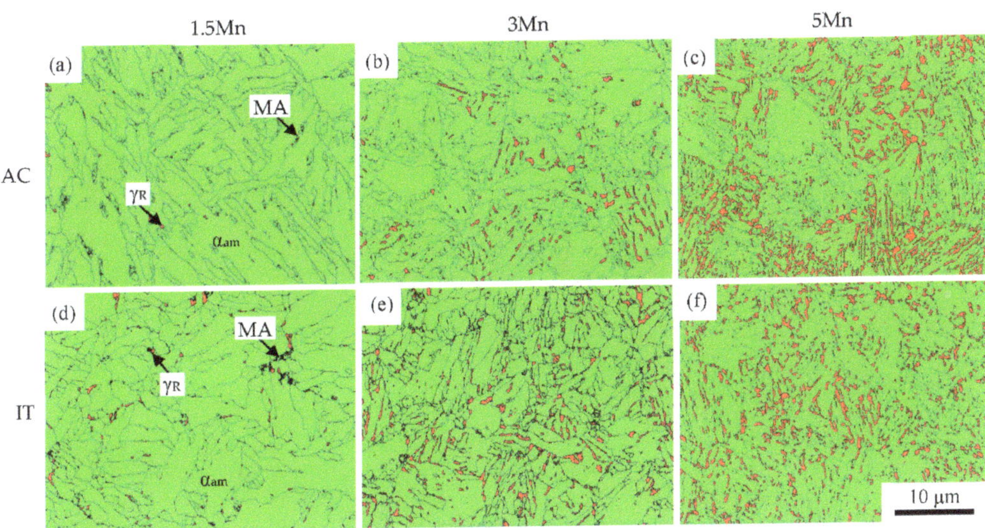

**Figure 3.** Phase maps of SEM-EBSD in the 1.5 Mn, 3 Mn and 5 Mn steels subjected to (**a–c**) the AC and (**d–f**) IT processes. Green and red phases denote annealed martensite ($\alpha_{am}$) and reverted austenite ($\gamma_R$), respectively. "MA" represents the martensite-austenite phase. Average lath sizes of annealed martensite in the 1.5 Mn, 3 Mn, and 5 Mn steels subjected to the AC and IT processes are 4.3 µm, 3.1 µm, and 2.4 µm, respectively.

**Table 2.** Reverted austenite characteristics, tensile properties at 25 °C and impact toughness properties of the 1.5 Mn, 3 Mn, and 5 Mn steels subjected to the AC and IT processes.

| Process | Steel | $f\gamma_0$ | $C\gamma_0$ | $k$ | YS | TS | UEl | TEl | RA | TS × TEl | $E_v$ (1) | $E_v$ (2) | DBTT |
|---|---|---|---|---|---|---|---|---|---|---|---|---|---|
| AC | 1.5 Mn | 14.1 | 0.72 | 10.3 | 415 | 924 | 25.7 | 33.8 | 48.1 | 31.2 | 123 | 132 | −30 |
| | 3 Mn | 24.8 | 0.55 | 23.4 | 320 | 1192 | 20.8 | 22.4 | 34.8 | 25.1 | 44 | 135 | 60 |
| | 5 Mn | 38.5 | 0.46 | 12.1 | 710 | 1256 | 32.0 | 36.8 | 29.7 | 46.2 | 64 | 148 | 35 |
| IT | 1.5 Mn | 11.4 | 0.95 | 5.0 | 500 | 875 | 22.2 | 32.4 | 57.3 | 28.3 | 175 | 178 | −65 |
| | 3 Mn | 17.3 | 0.68 | 15.3 | 495 | 977 | 21.4 | 28.3 | 59.6 | 27.6 | 105 | 153 | 10 |
| | 5 Mn | 36.2 | 0.52 | 9.4 | 680 | 1161 | 35.7 | 41.6 | 52.1 | 48.3 | 131 | 171 | −30 |

$f\gamma_0$ [vol. %]: initial volume fraction of reverted austenite, $C\gamma_0$ [mass %]: initial carbon concentration of reverted austenite, $k$: strain-induced transformation factor, YS [MPa]: yield stress or 0.2% offset proof stress, TS [MPa]: tensile strength, UEl [%]: uniform elongation, TEl [%]: total elongation, RA [%]: reduction of area, $E_v$ [J/cm$^2$]: Charpy V-notch impact value at 25 °C (1) and 100 °C (2), DBTT [°C]: ductile-brittle transition temperature.

Figure 5 shows the variations in reverted austenite fraction after cooling to −196 °C and heating to 300 °C in the 1.5 Mn, 3 Mn, and 5 Mn steels subjected to the AC and IT processes. The critical cooling temperature at which the reverted austenite fraction starts to decrease is lower than room temperature in both processes although the AC process raises slightly the critical cooling temperature. In the 1.5 Mn steel subjected to the IT process and the 3 Mn steel subjected to the AC process, the heating to temperatures above 150 °C and 250 °C decreases the reverted austenite fraction, respectively. According to Sugimoto et al. [36], the decrease in the reverted austenite fraction on cooling is owing to the ε-martensite transformation due to the decreased stacking fault energy of the reverted austenite [37,38]. On the other hand, the decrease in the reverted austenite fraction on heating may be caused by bainite transformation (or decomposition into ferrite and carbide) [35]. The decreases in the reverted austenite fraction by cooling in the 1.5 Mn, 3 Mn, and 5 Mn steels subjected to the AC process are larger than those subjected to the IT process. In both processes, the largest decrease in the reverted austenite fraction results in the 3 Mn steel.

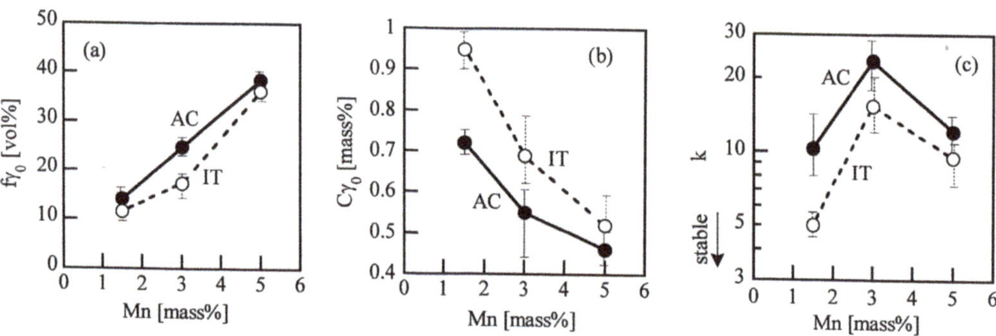

**Figure 4.** Variations in (**a**) initial volume fraction ($f_{\gamma 0}$), (**b**) initial carbon concentration ($C_{\gamma 0}$), and (**c**) strain-induced transformation factor ($k$) of reverted austenite as a function of Mn content in the steels subjected to the AC (●) and IT (○) processes after intercritical annealing.

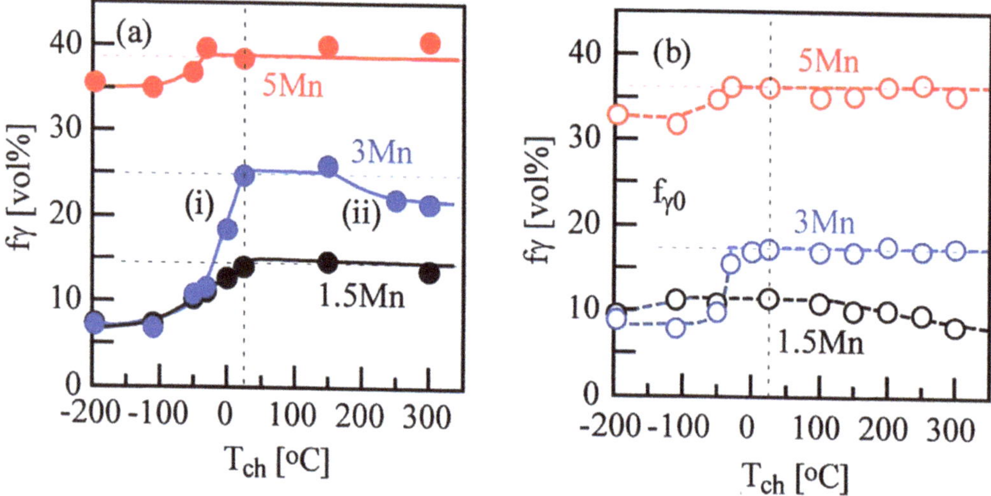

**Figure 5.** Variations in volume fraction of reverted austenite with cooling and heating temperature ($T_{ch}$) in the 1.5 Mn, 3 Mn, and 5 Mn steels subjected to (**a**) the AC and (**b**) IT processes. (i) and (ii) in (**a**) represent the decreases of reverted austenite by ε-martensite transformation [36] and bainite transformation (or decomposition into ferrite and carbide) [35], respectively. (**b**) is reprinted with permission from Wily-VCH GmbH, Weinheim: Steel Res. Int., Copyright 2021.

### 3.2. Tensile Properties at 25 °C

Typical engineering stress–strain curves at $T_t = 25$ °C in the 1.5 Mn, 3 Mn, and 5 Mn steels subjected to the AC and IT processes are shown in Figure 6. In all steels subjected to both processes, significant strain hardening takes place, although the yield plateau occurs in an early strain stage only in the 5 Mn steel. Moreover, many serrations occur only in the 5 Mn steel subjected to both processes.

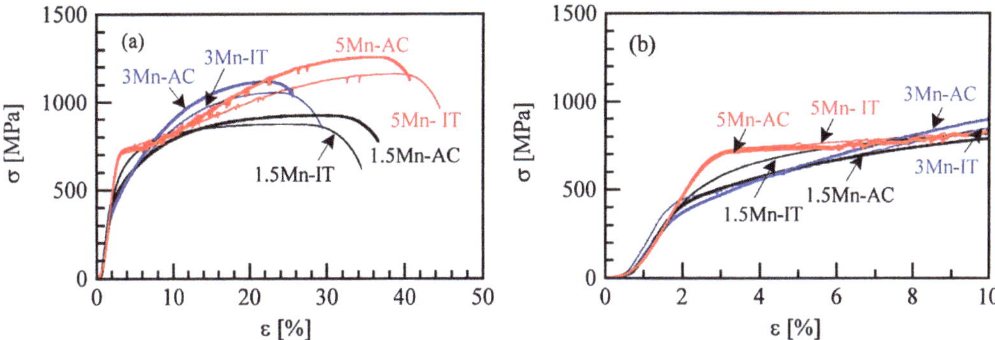

**Figure 6.** Engineering stress–strain ($\sigma$-$\varepsilon$) curves of (**a**) overall and (**b**) early strain ranges in the 1.5 Mn, 3 Mn, and 5 Mn steels subjected to the AC (thick lines) and IT (thin lines) processes.

Tensile properties of the corresponding steels are shown in Figure 7 and Table 2. The AC process increases the tensile strength compared to the IT process but decreases the yield stress or 0.2% offset proof stress except for the 5 Mn steel. In addition, the AC process brings on nearly the same TS × TEl as the IT process in all steels although it decreases the uniform and total elongations except for the 1.5 Mn steel.

In both processes, the 5 Mn steel exhibits the highest yield stress and tensile strength and the largest uniform and total elongations owing to the TRIP effect or SIMT hardening by a large amount of metastable reverted austenite and solid solution hardening by high Mn concentration, although the reduction of the area is the lowest [21]. The 5 Mn steel exhibits a superior TS × TEl (>45 GPa%). The 3 Mn steel possesses lower yield stress and higher tensile strength than the 1.5 Mn steel. The uniform and total elongations and reduction of the area are lower than those of the 1.5 Mn steel, despite a high reverted austenite fraction. The TS × TEl of the 3 Mn steel is lower than that of the 1.5 Mn steel.

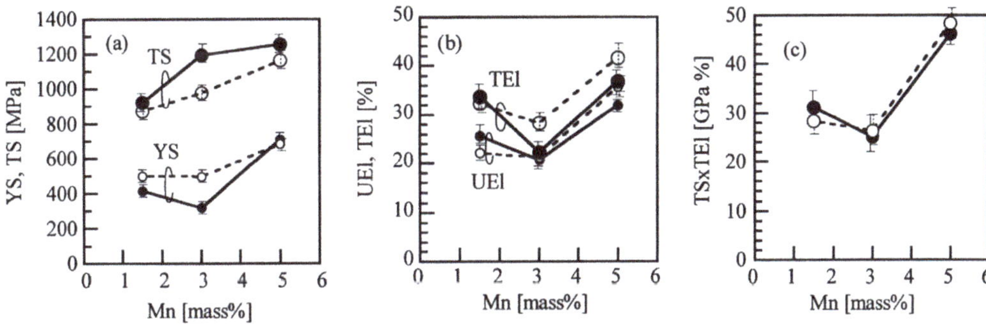

**Figure 7.** Variations in (**a**) yield stress or 0.2% offset proof stress (YS) and tensile strength (TS), (**b**) uniform and total elongations (UEl and TEl), and (**c**) the product of TS and TEl (TS × TEl) as a function of Mn content in the steels subjected to the AC (●) and IT (○) processes after intercritical annealing.

### 3.3. Impact Toughness

Figure 8a,b shows the Mn content dependences of the $E_v$ and the product of TS and $E_v$ (TS × $E_v$) at $T_t$ = 25 °C and 100 °C in the steels subjected to the AC and IT processes, respectively. The AC process deteriorates the $E_v$ and TS × $E_v$ at 25 °C in all steels, compared to the IT process. In both processes, the 3 Mn steel exhibits minimum $E_v$. The $E_v$ and TS × $E_v$ are considerably increased by 100 °C test with shorter error bars than them at 25 °C, especially in the 5 Mn steel subjected to the AC process. In this case, the $E_v$ of the 5 Mn steel is higher than that of the 1.5 Mn steel because an upper shelf $E_v$ appears at temperatures above 100 °C as mentioned later. The TS × $E_v$s at 100 °C of the 3 Mn and 5 Mn steels subjected to the AC and IT processes are nearly the same (Figure 8b).

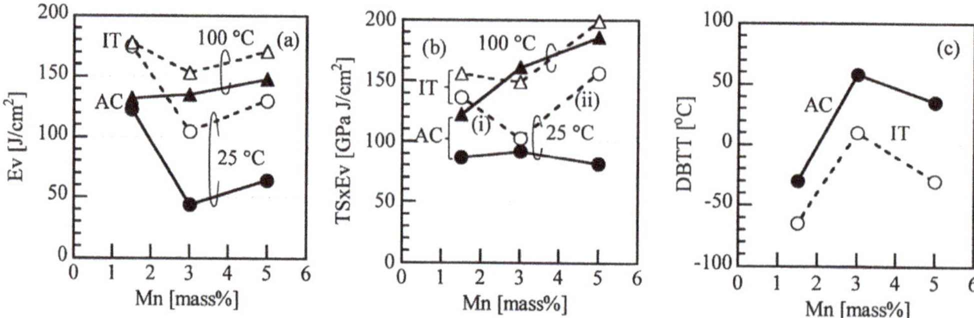

**Figure 8.** Variations in (**a**) Charpy V-notch impact value ($E_v$) and (**b**) the product of tensile strength and $E_v$ (TS × $E_v$) at $T_t$ = 25 °C (circle marks) and 100 °C (triangle marks) and (**c**) 50% shear fracture ductile-brittle transition temperature (DBTT) as a function of Mn content in the steels subjected to the AC (●▲) and IT (○△) processes. In (**b**), (i) and (ii) represent the contributions of mechanical stability and volume fraction of reverted austenite, respectively.

Figure 9 shows the impact load-displacement ($P$-$\delta$) curves at $T_t$ = 25 °C in the 1.5 Mn, 3 Mn, and 5 Mn steels subjected to the AC and IT processes. It is found that low $E_v$s of 1.5 Mn, 3 Mn, and 5 Mn steels subjected to the AC process mainly caused by low crack-propagation energy ($E_p$), although low $E_v$ of the 1.5 Mn steel subjected to the AC process is also associated with the low crack-initiation energy ($E_i$).

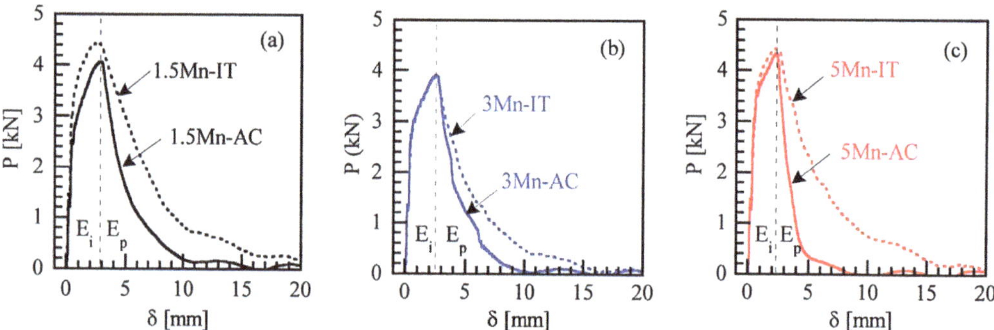

**Figure 9.** Impact load-displacement ($P$-$\delta$) curves at $T_t$ = 25 °C of (**a**) the 1.5 Mn, (**b**) 3 Mn, and (**c**) 5 Mn steels subjected to the AC (thick lines) and IT (thin lines) processes. $E_i$ and $E_p$ are crack-initiation and propagation energies, respectively, where $E_i$ + $E_p$ = $E_v$.

Figure 10a shows the relation between the $E_v$ and TS in the 1.5 Mn, 3 Mn, and 5 Mn steels subjected to the AC and IT processes which was tested at $T_t$ = 25 °C and 100 °C. At a testing temperature of 25 °C, the 5 Mn steel subjected to the AC process shows an equal $E_v$ and TS × $E_v$ to JIS-SCM420 steel quenched and tempered, and the 3 Mn steel exhibits lower $E_v$ and TS × $E_v$ than JIS-SCM420 steel. On the other hand, the $E_v$ and TS × $E_v$ at $T_t$ = 100 °C of the 5 Mn steel subjected to the AC and IT process significantly increase compared to those of JIS-SCM420 steel, in the same way as the 3 Mn steel. When compared to the $E_v$ of martensite-type 5 Mn steel, the TS × $E_v$s at $T_t$ = 100 °C of 5 Mn steel subjected to the AC and IT processes are higher than that at $T_t$ = 25 °C (151.4 GPa J/cm$^2$) of martensite-type 5 Mn steel [23].

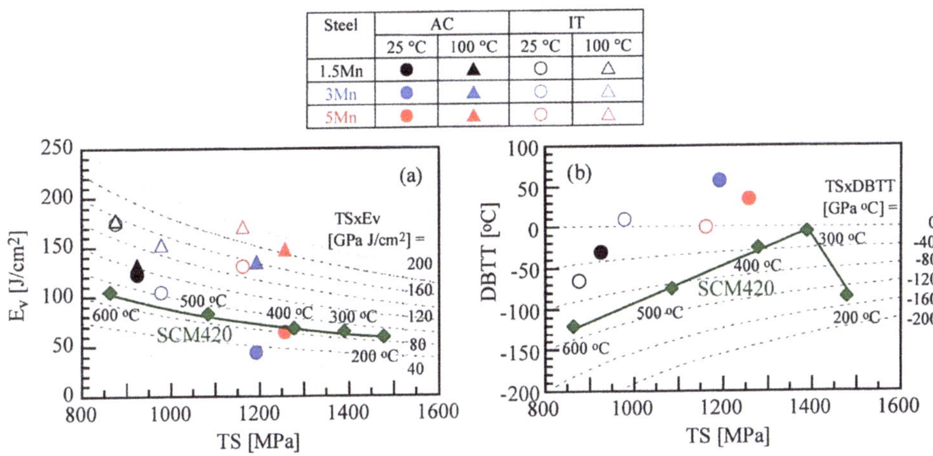

Figure 10. Variations in (**a**) Charpy V-notch impact value ($E_v$) at $T_t$ = 25 °C and 100 °C and (**b**) ductile-brittle transition temperature (DBTT) with tensile strength (TS) in the 1.5 Mn, 3 Mn, and 5 Mn steels subjected to the AC (solid marks) and IT (open marks) processes. ◆: $E_v$ at $T_t$ = 25 °C and DBTT of JIS-SCM420 steel (0.21%C-0.21%Si-0.77%Mn-1.0%Cr-0.2%Mo) quenched and tempered at 200 °C to 600 °C [7,21]. This figure is reproduced based on Ref. [21]. Copyright permission obtained.

Figure 11 shows the variations in the $E_v$ with the testing temperature of the 1.5 Mn, 3 Mn, and 5 Mn steels subjected to the AC and IT processes. Figure 10b shows the relation between the DBTT and TS of various steels. The AC process brings on higher DBTT than the IT process in the 1.5 Mn, 3 Mn, and 5 Mn steels. The DBTTs of 1.5 Mn, 3 Mn, and 5 Mn steels subjected to both processes are higher than those of quenched and tempered JIS-SCM420 steel. This result is very different from the result that the DBTTs of martensite-type 1.5 Mn, 3 Mn, and 5 Mn steels (−70 °C, −50 °C, and −50 °C, respectively, in a tensile strength range of 1274 MPa to 1633 MPa) are lower than those of the JIS-SCM420 steel [23]. It is noteworthy that the DBTTs of the 3 Mn and 5 Mn steels subjected to the AC process are higher than room temperature (25 °C). In both processes, the 1.5 Mn and 3 Mn steels exhibit the lowest and highest DBTTs, respectively.

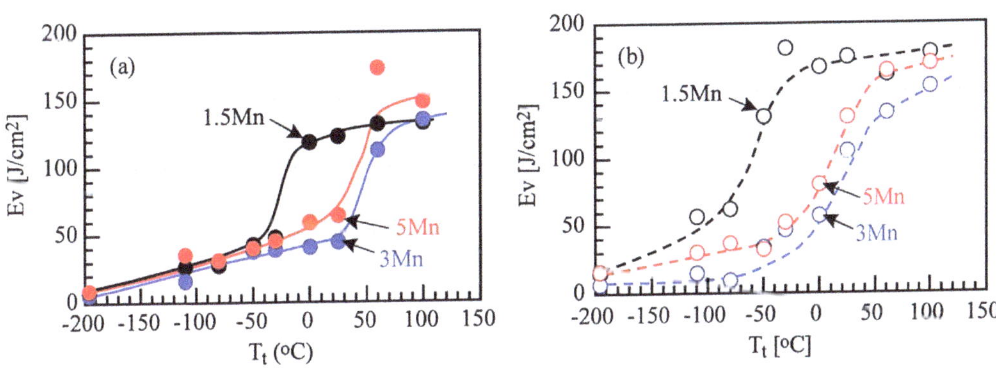

Figure 11. Variations in Charpy V-notch impact value ($E_v$) with impact test temperature ($T_t$) in the 1.5 Mn, 3 Mn, and 5 Mn steels subjected to (**a**) the AC and (**b**) IT processes. (**b**) is reprinted with permission from Spring Nature: Metall. Mater. Trans. A, Copyright 2021.

Figure 12 shows SEM images of fracture surfaces of the 1.5 Mn, 3 Mn, and 5 Mn steels subjected to the AC process after impact tests at 25 °C and −196 °C. In Figure 12a–c, the fracture of all steels appears as dimple one. The fracture surface of the 3 Mn and 5 Mn steels is characterized by uniform fine dimples, different from that of 1.5% Mn steel consisting of both coarse and fine dimples. In Figure 12d–f, conventional cleavage fracture is observed in the 1.5 Mn steel fractured at −196 °C, but non-quasi-cleavage fracture appears on the surface of the 5 Mn steel although a small amount of cleavage facet coexists with the non-quasi-cleavage fracture. The 3 Mn steel shows a similar fracture surface to 5 Mn steel. The above-mentioned non-quasi-cleavage fracture looks like an inter-granular fracture along prior austenitic grain boundary which appeared in the 5 Mn steel subjected to the IT process [21].

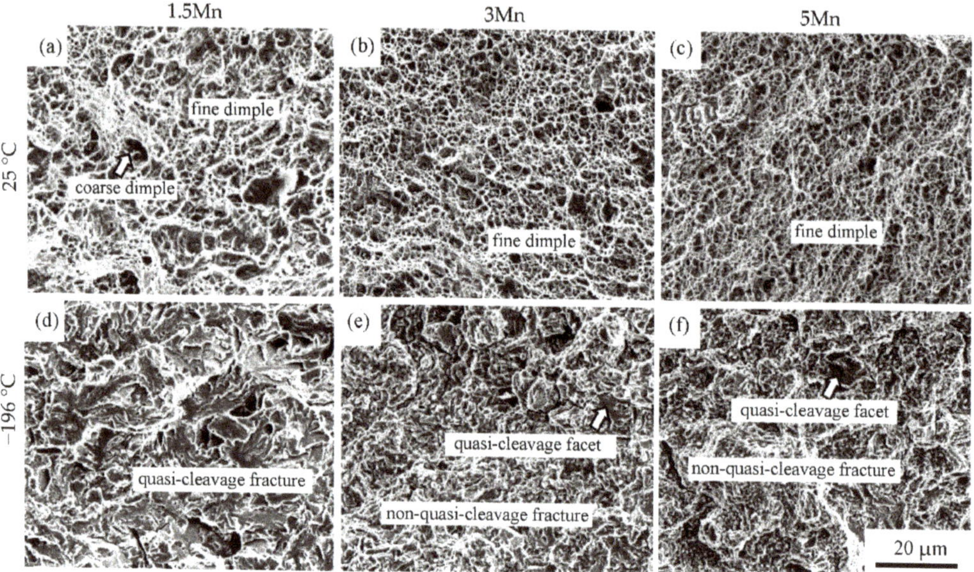


**Figure 12.** SEM images of fracture surface in the 1.5 Mn, 3 Mn, and 5 Mn steels subjected to the AC process after impact tests at (**a–c**) 25 °C and (**d–f**) −196 °C.

## 4. Discussion

### 4.1. $E_v$ at 25 °C and 100 °C

The 5 Mn steel subjected to the IT process possessed lower $E_v$ and TS × $E_v$ at 25 °C than the 1.5 Mn steel subjected to the same process (Figure 8a,b). According to Sugimoto et al. [21], they are essentially caused by the SIMT (or TRIP effect) of a large amount of reverted austenite, solid solution hardening of the matrix structure by high solute Mn concentration and microstructure-refining, although low mechanical stability of the reverted austenite plays a role in reducing these values. The SIMT suppresses the void or dimple formation through the plastic relaxation of localized stress concentration by volume change from the reverted austenite to martensite. The present 3 Mn steel subjected to the IT process exhibited the minimum $E_v$ and TS × $E_v$ (Figure 8a,b). Sugimoto et al. [21] showed that this was mainly associated with lower mechanical stability than that of the 5 Mn steel (Figure 4b,c), as well as lower volume fraction (Figure 4c) and larger inter-particle path of reverted austenite. In the 1.5 Mn steel, the MA phase plays a role in forming coarse dimples and resultantly increases the $E_v$, as well as the SIMT of the reverted austenite [21,39]. The 3 Mn and 5 Mn steels subjected to the IT process showed uniform fine dimple fracture on impact test at 25 °C. In this case, the dimples were mainly initiated at the reverted austenite (or the strain-induced martensite)/matrix structure interface [21].

In the present study, the AC process decreased the $E_v$ at 25 °C in the 5 Mn steel, compared to the IT process, in the same way as the 1.5 Mn and 3 Mn steels (Figure 8a). In all steels, dimple fracture was observed, although the mixed coarse and fine dimples were formed only in the 1.5 Mn steel (Figure 12a–c). Figure 9c showed that the decrease in the $E_v$ was mainly caused by low crack-propagation energy. As shown in Figure 4, the AC process increased the reverted austenite fraction but decreased the mechanical stability of the reverted austenite (increased the $k$-value). As the other microstructures were nearly the same between the steels subjected to the AC and IT processes, the decreased mechanical stability is considered to mainly reduce the $E_v$ due to the easy SIMT which promotes easy void formation at the strain-induced martensite/annealed martensite interface and the resultant easy void propagation and connection. High $E_v$ of the 1.5 Mn steel subjected to the AC process may be caused by a small amount of MA phase like in the case of the IT process.

As shown in Figure 8a, Figure 10a, and Figure 11, the impact test at 100 °C significantly increased the $E_v$ and TS × $E_v$ in the 3 Mn and 5 Mn steels subjected to the AC process, compared to the IT process. Sugimoto et al. [36] investigated the testing temperature dependence of the mechanical stability of reverted austenite in the 1.5 Mn, 3 Mn, and 5 Mn steels subjected to the IT process and found that the mechanical stability of the 3 Mn and 5 Mn steels is considerably influenced by the testing temperature compared to that of the 1.5 Mn steel as shown in Figure 13. Namely, the mechanical stability at the temperatures lower than 25 °C is lower than that of the 1.5 Mn steel but at temperatures between 100 °C and 200 °C it is higher than that of 1.5 Mn steel. If a similar testing temperature dependence of the $k$-value is obtained in the AC process, although the $k$-values are higher than those of the IT process because of lower carbon concentration (Figure 4b), the increased mechanical stability of reverted austenite is considered to increase the $E_v$s and TS × $E_v$s at 100 °C in the 3 Mn and 5 Mn steels subjected to the AC process.

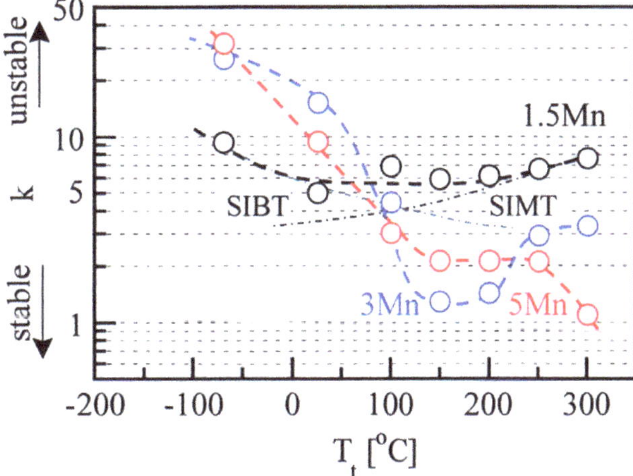


**Figure 13.** Variations in $k$-value with tensile test temperature ($T_t$) in the 1.5 Mn, 3 Mn and 5 Mn steels subjected to the IT process [36]. "SIMT" and "SIBT" represent the strain-induced martensite and bainite transformation, respectively. Reprinted with permission from Wily-VCH GmbH, Weinheim: Steel Res. Int., Copyright 2021.

*4.2. DBTT*

In general, Mn plays a role in increasing the DBTT due to Mn segregation on the prior austenitic grain boundary [40–42]. According to Yamanaka and Kowaka [43], Mn addition below 2 mass % lowers the DBTT but Mn addition above 5 mass % raises the DBTT in 0.002%C-(1−7.5)%Mn ferritic steels. In the latter, inter-granular fracture occurred at a low-temperature range, not cleavage fracture. Tanaka et al. [44] investigated the effect of Mn addition on the DBTT in 0.002%C-(0−2)%Mn-0.03%Ti ferritic steels and showed that Mn raises the DBTT by lowering the surface energy for inter-granular fracture. Based on the above results and SEM observation of the fracture surface, Sugimoto et al. [21] reported that high DBTTs of the 3 Mn and 5 Mn steels subjected to the IT process are associated with the inter-granular fracture in a brittle fracture temperature range and non-quasi-cleavage fracture (inter-granular fracture) (Figure 14b), although the quasi-cleavage fracture predominantly occurs in the 1.5 Mn steel with low DBTT (Figure 14a). They also proposed that the inter-granular fracture resulted from (i) high Mn segregation and (ii) a large amount of unstable reverted austenite (or strain-induced martensite) on the prior austenitic grain boundaries. Tanaka et al. [44] also reported that solute Mn decreases the activation energy of dislocation glide at low temperatures like Ni and suppresses the cleavage fracture. According to Sugimoto et al. [23], quasi-cleavage fracture occurred at −196 °C in the martensite-type 5 Mn steel with a small amount of retained austenite (7.6 vol. %). This indicates that a large amount of reverted austenite also suppresses the cleavage fracture and promotes the inter-granular fracture in the present 5 Mn steel subjected to the AC and IT processes.

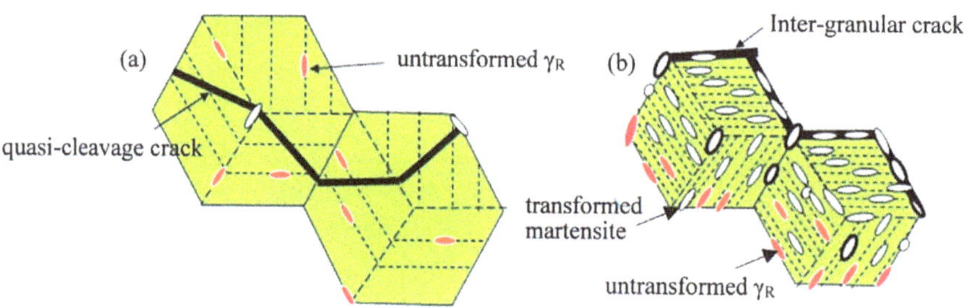


**Figure 14.** Brittle fracture morphology at −196 °C appeared on the fracture surface of impact tested samples of (**a**) the 1.5 Mn and (**b**) 5 Mn steels subjected to the AC and IT processes. The 3 Mn steel shows the intermediate fracture mode in which quasi-cleavage fracture and inter-granular fracture are coexisting.

According to Kunitake et al. [45], the DBTT of 0.12%C-0.30%Si-0.83%Mn-0.30%Cu-1.11%Ni-0.53%Cr-0.49%Mo-0.03%V steel with the microstructure of martensite and martensite/lower bainite structures correlates with the unit crack path ($d_p$) on the quasi-cleavage fracture surfaces as given by the following equation,

$$\text{DBTT} = \text{E} - \text{Z} \times \log d_p^{-1/2}. \tag{2}$$

where E and Z are the material's constants. The unit crack path is corresponding to the packet size. Figure 15 shows the relation between the DBTT and $d_p$ for various steels. The DBTTs of quenched and then tempered JIS-SCM420 steel are also plotted in the figure. In this case, the grain facet sizes of the 3 Mn and 5 Mn steels were assumed to be the same as the $d_p$. If the slopes of these DBTT versus $\log d_p^{-1/2}$ lines are assumed to be similar to that reported by Kunitake et al. [45], the 3 Mn and 5 Mn steels subjected to the IT process had higher DBTTs by 128 °C ($\Delta$DBTT$_1$) than the 1.5 Mn steel and JIS-SCM420 steel. This means that the $d_p$ of cleavage fracture is equivalent to grain facet size in the 5 Mn steel and the $\Delta$DBTT$_1$ is associated with the inter-granular fracture.

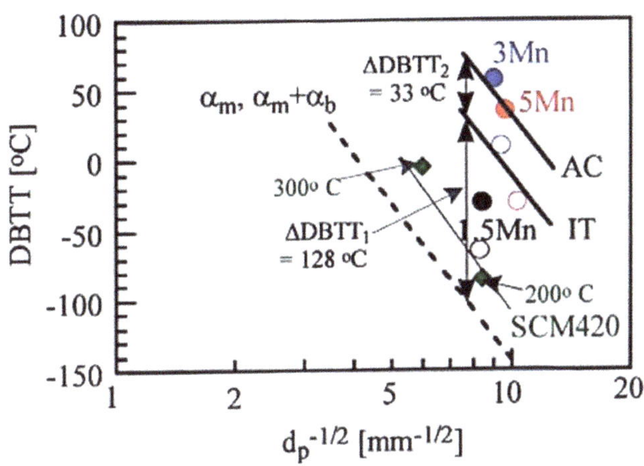



**Figure 15.** Relationship between 50% shear fraction ductile-brittle transition temperature (DBTT) and unit crack path ($d_p$) for the 1.5 Mn (●○), 3 Mn (●○), and 5 Mn (●○) steels subject to the AC (solid marks) and IT (open marks) processes and JIS-SCM420 (◆) steel quenched and then tempered at 200 °C and 300 °C. A dotted line represents the DBTT of 0.12%C-0.30%Si-0.83%Mn-0.30%Cu-1.11%Ni-0.53%Cr-0.49%Mo-0.03%V steel with the microstructure of martensitic ($\alpha_m$) or martensitic/bainitic ($\alpha_m + \alpha_b$) structures [45]. This figure is reproduced based on Ref. [21]. Copyright obtained.

The AC process further raised the DBTTs of the 3 Mn and 5 Mn steels by 33 °C ($\Delta$DBTT$_2$), compared to the IT process (see Figure 15). The AC process increased the reverted austenite fraction but decreased the thermal and mechanical stability, compared to the IT process in the 1.5 Mn, 3 Mn, and 5 Mn steels. Because the AC process produced the same microstructure as the IT process except for the reverted austenite characteristics, an increase in the DBTT ($\Delta$DBTT$_2$) of the 3 Mn and 5 Mn steels subjected to the AC process may be mainly caused by the further decrease in the mechanical stability of reverted austenite.

In this case, the inter-granular fracture raises the DBTT by $\Delta$DBTT$_1$ in the same way as the IT process.

## 5. Conclusions

The reverted austenite characteristics, tensile properties, and impact toughness of 0.2%C–1.5%Si–5.0%Mn D-MMn steel subjected to the AC process were compared with those subjected to the IT process. The main results obtained are as follows:

(1)  The AC process increased the volume fraction of reverted austenite in the annealed martensite matrix structure of the 5 Mn steel, but decreased the thermal and mechanical stability, compared to the IT process.

(2)  The AC process increased the tensile strength but decreased the total elongation. The TS × TEl was hardly influenced by the cooling process routs.

(3)  The TS × $E_v$ at 25 °C of the 5 Mn steel was deteriorated by the AC process compared to the IT process, although at 100 °C it improved by the AC process in the same way as the IT process. The deterioration of the impact toughness at 25 °C was mainly caused by the decreased mechanical stability of reverted austenite, although the increased volume fraction of metastable reverted austenite contributed to improving the impact toughness.

(4)  The AC process further raised the DBTT of the 5 Mn steel compared to the IT process. The high DBTT was mainly associated with the further decrease in the thermal and mechanical stability of the reverted austenite, as well as the inter-granular fracture in a brittle fracture temperature range in a similar way as the IT process.

**Author Contributions:** The first draft of the paper was written by K.-i.S., and the final editing was done by K.-i.S. and H.T. All authors have read and agreed to the published version of the manuscript.

**Funding:** This research received no external funding.

**Institutional Review Board Statement:** Not applicable.

**Informed Consent Statement:** Not applicable.

**Data Availability Statement:** Not applicable.

**Acknowledgments:** I thank Junya Kobayashi from Ibaraki University for the kind discussion.

**Conflicts of Interest:** The author declares no conflict of interest.

# References


1. Rana, R.; Singh, S.B. *Automotive Steels—Design, Metallurgy, Processing and Applications*; Woodhead Publishing: Cambridge, UK, 2016.
2. Bleck, W.; Guo, X.; Ma, Y. The TRIP effect and its application in cold formable sheet steels. *Steel Res. Int.* **2017**, *88*, 1700218. [CrossRef]
3. Ghosh, M.; Bansal, G.; Chandan, A.; Shah, M.; Tripathy, S.; Murugaiyan, P.; Sahoo, B.; Mukherjee, K.; Srivastava, V.C.; Chowdhury, S.G. Functionally driven steels: A review. *Steel Tech.* **2017**, *12*, 13–36.
4. Sugimoto, K.; Hojo, T.; Srivastava, A.K. Low and medium carbon advanced high-strength forging steels for automotive applications. *Metals* **2019**, *9*, 1263. [CrossRef]
5. Tong, C.; Rong, Q.; Yardley, V.; Li, X.; Luo, J.; Zhu, G.; Shi, Z. New developments and future trends in low-temperature hot stamping technologies: A review. *Metals* **2020**, *10*, 1652. [CrossRef]
6. Sugimoto, K.; Hojo, T.; Mizuno, Y. Effects of fine particle peening conditions on the rotational bending fatigue strength of a vacuum-carburized transformation-induced plasticity-aided martensitic Steel. *Metall. Mater. Trans. A* **2018**, *49A*, 1552–1560. [CrossRef]
7. Sugimoto, K. Recent progress of low and medium carbon advanced martensitic steels. *Metals* **2021**, *11*, 652. [CrossRef]
8. Sugimoto, K.; Tsunezawa, M.; Hojo, T.; Ikeda, S. Ductility of 0.1-0.6C-1.5Si-1.5Mn ultra high-strength TRIP-aided sheet steels with bainitic ferrite matrix. *ISIJ Int.* **2004**, *44*, 1608–1614. [CrossRef]
9. Sugimoto, K.; Murata, M.; Song, S. Formability of Al-Nb bearing ultrahigh-strength TRIP-aided sheet steels with bainitic ferrite and/or martensite matrix. *ISIJ Int.* **2010**, *50*, 162–168. [CrossRef]
10. Sugimoto, K.; Kobayashi, J.; Pham, D.V. Advanced ultrahigh-strength TRIP-aided martensitic sheet steels for automotive applications. In Proceedings of the New Developments in Advanced High Strength Sheet Steels (AIST 2013), Vail, CO, USA, 23–27 June 2013; pp. 175–184.
11. Pham, D.V.; Kobayashi, J.; Sugimoto, K. Effects of microalloying on stretch-flangeability of TRIP-aided martensitic sheet steel. *ISIJ Int.* **2014**, *54*, 1943–1951. [CrossRef]
12. Tian, J.; Xu, G.; Zhou, M.; Hu, H. Refined bainite microstructure and mechanical properties of a high-strength low-carbon bainitic steel. *Steel Res. Int.* **2018**, *89*, 1700469. [CrossRef]
13. Navarro-López, A.; Hidalgo, J.; Sietsma, J.; Santofimia, M.J. Influence of the prior athermal martensite on the mechanical response of advanced bainitic steel. *Mater. Sci. Eng. A* **2018**, *735*, 343–353. [CrossRef]
14. Speer, J.G.; De Moor, E.; Findley, K.O.; Matlock, B.C.; De Cooman, B.C.; Edmonds, D.V. Analysis of microstructure evolution in quenching and partitioning automotive sheet steel. *Metall. Mater. Trans. A* **2011**, *42A*, 3591–3601. [CrossRef]
15. Tan, X.; Xu, Y.; Yang, X.; Wu, D. Microstructure–properties relationship in a one-step quenched and partitioned steel. *Mater. Sci. Eng. A* **2014**, *589*, 101–111. [CrossRef]
16. Miller, R.I. Ultrafine-grained microstructures and mechanical properties of alloy steels. *Metall. Trans.* **1972**, *3*, 905–912. [CrossRef]
17. Furukawa, T. Dependence of strength-ductility characteristics on thermal history in low carbon, 5 wt-%Mn steels. *Mater. Sci. Technol.* **1989**, *5*, 465–470. [CrossRef]
18. Hu, B.; He, B.; Cheng, G.; Yen, H.; Huang, M.; Luo, H. Super-high-strength and formable medium Mn steel manufactured by warm rolling process. *Acta Mater.* **2019**, *174*, 131–141. [CrossRef]
19. Su, G.; Gao, X.; Zhang, D.; Du, L.; Hu, J.; Liu, Z. Impact of reverted austenite on the impact toughness of the high-strength steel of low carbon medium manganese. *JOM* **2018**, *70*, 672–679. [CrossRef]
20. Cao, W.; Wang, C.; Shi, J.; Wang, M.; Hui, W.; Dong, D. Microstructure and mechanical properties of Fe–0.2C–5Mn steel processed by ART-annealing. *Mater. Sci. Eng. A* **2011**, *528*, 6661–6666. [CrossRef]
21. Tanino, H.; Horita, M.; Sugimoto, K. Impact toughness of 0.2 pct-1.5 pct-(1.5 to 5) pct Mn transformation-induced plasticity-aided steels with an annealed martensite matrix. *Metall. Mater. Trans. A* **2016**, *47A*, 2073–2080. [CrossRef]
22. Hanamura, T.; Torizuka, S.; Tamura, S.; Enokida, S.; Takechi, H. Effect of austenite grain size on transformation behavior, microstructure and mechanical properties of 0.1C-5Mn martensitic steel. *ISIJ Int.* **2013**, *53*, 2218–2225. [CrossRef]
23. Sugimoto, K.; Tanino, H.; Kobayashi, J. Impact toughness of medium-Mn transformation-induced plasticity-aided steels. *Steel Res. Int.* **2015**, *86*, 1151–1160. [CrossRef]

24. Maeda, T.; Okuhara, S.; Matsuda, K.; Matsumura, T.; Tsuchiyama, T.; Shirahata, H.; Kawamoto, Y.; Fujioka, M. Toughening mechanism in 5%Mn and 10%Mn martensitic steels treated by thermo-mechanical control process. *Mater. Sci. Eng. A* **2021**, *812*, 141058. [CrossRef]
25. Cao, W.; Zhang, M.; Huang, C.; Xiao, S.; Dong, H.; Weng, Y. Ultrahigh Charpy impact toughness (-450J) achieved in high strength ferrite/martensite laminated steels. *Sci. Rep.* **2017**, *7*, 41459. [CrossRef]
26. Liu, L.; Yu, Q.; Wang, Z.; Ell, J.; Huang, M.X.; Ritchie, R.O. Making ultrastrong steel tough by grain-boundary delamination. *Science* **2020**, *368*, 1347–1352. [CrossRef]
27. Zou, Y.; Xu, Y.; Hu, Z.; Gu, X.; Peng, F.; Tan, X.; Chen, S.; Han, D.; Misra, R.D.K.; Wang, G. Austenite stability and its effect on the toughness of a high strength ultra-low carbon medium manganese steel plate. *Mater. Sci. Eng. A* **2016**, *675*, 153–163. [CrossRef]
28. Qi, X.; Du, L.; Hu, J.; Misra, R.D.K. Effect of austenite stability on toughness, ductility, and work-hardening of medium-Mn steel. *Mater. Sci. Technol.* **2019**, *35*, 2134–2142. [CrossRef]
29. Su, G.; Gao, X.; Yan, T.; Zhang, D.; Cui, C.; Du, L.; Liu, Z.; Tang, Y.; Hu, J. Intercritical tempering enables nanoscale austenite/ε-martensite formation in low-C medium-Mn steel: A pathway to control mechanical properties. *Mater. Sci. Eng. A* **2018**, *736*, 417–430. [CrossRef]
30. Torizzo, Q.; Maziere, M.; Perlade, A.; Gourgues-Lorenzon, A.-F. Effect of austenite stability on the fracture micromechanisms and ductile-to-brittle transition in a medium-Mn, ultra-fine-grained for automotive applications. *J. Mater. Sci.* **2020**, *55*, 9245–9257.
31. Gramlich, A.; Emmrich, R.; Bleck, W. Austenite reversion tempering-annealing of 4 wt.% manganese steels for automotive forging application. *Metals* **2019**, *9*, 575. [CrossRef]
32. Chang, Y.; Wang, C.; Zhao, K.; Dong, H.; Yan, J. An introduction to medium-Mn steel: Metallurgy, mechanical properties and warm stamping process. *Mater. Des.* **2016**, *94*, 424–432. [CrossRef]
33. Maruyama, H. X-ray measurement of retained austenite. *Jpn. Soc. Heat Treat.* **1977**, *17*, 198–204.
34. Dyson, D.J.; Holmes, B. Effect of alloying additions on the lattice parameter of austenite. *J. Iron Steel Inst.* **1970**, *208*, 469–474.
35. Sugimoto, K.; Usui, N.; Kobayashi, M.; Hashimoto, S. Effects of volume fraction and stability of retained austenite on ductility of TRIP-aided dual-phase steels. *ISIJ Int.* **1992**, *32*, 1311–1318. [CrossRef]
36. Sugimoto, K.; Hidaka, S.; Tanino, H.; Kobayashi, J. Effects of Mn content on the warm stretch-flangeability of C-S-Mn TRIP-aided steels. *Steel Res. Int.* **2017**, *83*, 1600482. [CrossRef]
37. Zambrano, O.A. Stacking fault energy maps of Fe-Mn-Al-C-Si steels: Effect of temperature, grain size and variations in compositions. *J. Eng. Mater. Technol.* **2016**, *138*, 041010. [CrossRef]
38. Nakano, J.; Jacques, P.J. Effects of the thermodynamic parameters of the hcp phase on the stacking fault energy calculations in the Fe-Mn and Fe-Mn-C systems. *Calphad* **2010**, *34*, 167–175. [CrossRef]
39. Kobayashi, J.; Ina, D.; Nakajima, Y.; Sugimoto, K. Effects of microalloying on the impact toughness of ultrahigh-strength TRIP-aided martensitic steels. *Metall. Mater. Trams. A* **2013**, *44A*, 5006–5017. [CrossRef]
40. Nasim, M.; Edwards, B.C.; Wilson, E.A. A study of grain boundary embrittlement in an Fe-8%Mn alloy. *Mater. Sci. Eng. A* **2000**, *281*, 56–67. [CrossRef]
41. Heo, N.; Nam, J.; Heo, Y.; Kim, S. Grain boundary embrittlement by Mn and eutectoid reaction in binary Fe-12Mn steel. *Acta Mater.* **2013**, *61*, 4022–4034. [CrossRef]
42. Kuzmina, M.; Ponge, D.; Raabe, D. Grain boundary segregation engineering and austenite reversion turn embrittlement into toughness: Example of a 9wt.% medium Mn steel. *Acta Mater.* **2015**, *86*, 182–192. [CrossRef]
43. Yamanaka, K.; Kowaka, M. Mechanical properties and fracture behavior of Fe-Mn alloys. *J. Jpn. Inst. Met.* **1979**, *43*, 1151–1157. [CrossRef]
44. Tanaka, M.; Maeno, K.; Yoshimura, N.; Hoshino, M.; Uemori, R.; Ushioda, K.; Higashida, K. Effect of Mn addition on a brittle-to-ductile transition in ferritic steels. *Tetsu Hagane* **2014**, *100*, 1267–1273. [CrossRef]
45. Kunitake, T.; Terasaki, F.; Ohmori, Y.; Ohtani, H. *The Effect of Transformation Structures on the Toughness of Quenched-and-Tempered Low-Carbon Low-Alloy Steels. Toward Improved Ductility and Toughness*; Climax Molybdenum Development Company Ltd.: Kyoto, Japan, 1971; pp. 83–100. (In Japanese)

*Article*

# Optimization of Open Die Ironing Process through Artificial Neural Network for Rapid Process Simulation

**Silvia Mancini** [1], **Luigi Langellotto** [1], **Giovanni Zangari** [1], **Riccardo Maccaglia** [2] and **Andrea Di Schino** [2,*]

1   RINA Consulting Centro Sviluppo Materiali, Via di Castel Romano 100, 00128 Roma, Italy; silvia.mancini@rina.org (S.M.); luigi.langellotto@rina.org (L.L.); giovanni.zangari@rina.org (G.Z.)
2   Dipartimento di Ingegneria, Università degli Studi di Perugia, Via G. Duranti 93, 06125 Perugia, Italy; riccardo.maccaglia@studenti.unipg.it
*   Correspondence: andrea.dischino@unipg.it

Received: 24 September 2020; Accepted: 18 October 2020; Published: 21 October 2020

**Abstract:** The open die forging sequence design and optimization are usually performed by simulating many different configurations corresponding to different forging strategies. Finite element analysis (FEM) is a tool able to simulate the open die forging process. However, FEM is relatively slow and therefore it is not suitable for the rapid design of online forging processes. A new approach is proposed in this work in order to describe the plastic strain at the core of the piece. FEM takes into account the plastic deformation at the core of the forged pieces. At the first stage, a thermomechanical FEM model was implemented in the MSC.Marc commercial code in order to simulate the open die forging process. Starting from the results obtained through FEM simulations, a set of equations describing the plastic strain at the core of the piece have been identified depending on forging parameters (such as length of the contact surface between tools and ingot, tool's connection radius, and reduction of the piece height after the forging pass). An Artificial Neural Network (ANN) was trained and tested in order to correlate the equation coefficients with the forging to obtain the behavior of plastic strain at the core of the piece.

**Keywords:** open die forging; artificial neural network; fast simulator; process optimization

---

## 1. Introduction

Forged steels represent a quite interesting material family, both from a scientific and commercial point of view, following many applications they can be devoted to [1]. Moreover, it is essential to deeply understand the relations between properties and microstructure and how to drive them by process [2–6]. Such steels are widely applied in the machining and forming industry [7–9], in automotive applications [10], and in other fields including aerospace, transport, and precision industries [11–13]. Moreover, along with the development of the energy industry and the growing power of power engineering devices, the demand for large-sized hot forged products has also increased. This includes turbine shafts (water, gas, steam), rotors for wind, and gas power generators [14]. The forecast of the Forging Industry Association (FIA) regarding the steady increase in demand for forgings used in the power engineering and oil industries was confirmed at the 20th International Forgemasters Meeting (IFM'2017) in Austria. Furthermore, based on data from EUROFORGE (an organization that associates European production associations, including the Polish Forge Association), the volume of forged products has been growing steadily, and in 2020 it will reach over 10 million tons. The global forging market is likely to grow significantly at a compound annual growth rate (CAGR) of close to 8%, reaching USD 111.1 billion by 2020, according to Technavio's latest report [15].

In all the above applications, forgings are adopted following a requirement in terms of increasing manufacturing economic efficiency and enhancing mechanical properties, such as high strength, wear resistance, hardness, and toughness [16,17].

Free forging is the oldest forging method commonly adopted to forge heavy charge materials in short production runs. High-pressure hydraulic forging presses are on the other hand used in open die forging of heavy steel forgings (carbon, alloy, high-alloy, stainless and other steels). Open die forging is an incremental forging process that is mainly used to produce large parts requiring high mechanical properties and reliability of the forged parts [16]. In the steel industry, there is a need to produce several large components characterized by high weight, which require high press loads. Such components can be spindles, rolls for rolling mills as well large turbine shafts and nuclear reactor vessels [18,19]. In the open die forging process, the workpiece is processed using dies, which can be flat or shaped (e.g., concave or V-shape). The piece undergoes a plastic deformation at high temperature when is pressed by a series of multiple strokes along the feed direction. In this way the piece changes both its geometry and internal properties [20].

The forging process is able to reduce many defects induced by casting process (e.g., cavities, porosities), thus allowing to manufacture almost defect free pieces by assuring a homogeneous plastic strain in the workpiece [21,22]. The quality of the open die forging process depends on several parameters such as die width, the shape of the die, die overlap, die stagger, ingot shape and dimensions, temperature gradient, pass schedule, and so on [23]. In order to achieve the requirements in term of geometric tolerance and internal quality, it is common practice to set up an appropriate pass schedule which has to be verified using numerical simulations. In order to do that, in the case of forging sequence design and optimization, it is necessary to simulate many different configuration corresponding to different forging strategies in order to identify the best solution. FEM is one of the most commonly adopted approach, anyway, it is reported that it requires significant efforts in terms of both computational resources and time [24,25].

The second limitation of the FEM approach in designing the forging strategy: in fact, it is not always possible to predict process conditions because the manufacturing processes are often quite unpredictable if compared with pilot or laboratory process, as often adopted to validate the simulate. For this reason, it is useful to develop fast calculation models of the open die forging process that allow to perform a rapid calculation of material properties during the layout of the process as well for the online monitoring of the process.

In the past, some authors dedicated to the development of fast models oriented to open die forging process optimization [26–30]. The common idea at the base of the above works was to develop process models able to combine data from online measurements and a simplified plasto-mechanical model for the forecasting of the equivalent strain, strain rate, and the temperature in the core of the forged piece. The final aim of such models is to optimize the stretching forging not only from a geometric point of view but also in terms of final microstructure, internal quality (e.g., casting porosity closure), working temperature to avoid phase transformation during the mechanical processing. Kim et al. [27] developed forging pass schedule algorithms based on artificial neural networks (ANN) are mainly oriented to calculate the optimum number of passes and reduction in each pass to economize power and minimize the forging cycle time. The algorithms were trained on the experimental data from pilot and full-scale industrial forging.

Starting from the approach reported in [28–30], a new formulation based on artificial neural networks (ANN) [31,32] is proposed in this work to quickly (fraction of a second) and correctly evaluate the plastic strain at the core of the piece. Starting from the coefficients for the new analytical model, a neural network has been implemented and trained using analytic coefficients in order to obtain a fast estimation of the plastic strain that is able to allow rapid and accurate evaluation of deformation occurring during the forging process. The correct evaluation of plastic strain at the core fiber of the forged product is a parameter to consider in a tool that aims at the internal integrity of the material or the final microstructure optimization.

## 2. Methods

### 2.1. FEM Model for Open-Die Forging of 42crmo4 Steel Grade

The open die forging of large products is a very complex process made of a sequence of several forging operations like upsetting, cogging, drawing, also including the furnace soaking to reheat the pieces between deformation steps. In order to fulfill the final requirements for the product, the understanding and prediction of microstructures that develop during deformation processing are necessary. In the stretching forging process, the pass sequence is roughly square or round to octagon to round, where the reduction ratio only varies every second pass. In the considered pass sequence, the round is forged to an oval section with a given height reduction, turned 90°, and then forged again with the same reduction and same bite ratio, producing a square cross-section. The newly obtained square bar is then forged into an octagonal bar, which is an intermediate shape between square and round. Finally, the round bar is produced in successive passes by deforming the octagonal bar. The octagonal bar has a greater cross-sectional area than the final round bar. The round bar is finished in a round-contoured die during the finishing passes. Figure 1 shows the overall forging sequence from a square billet to a round bar.

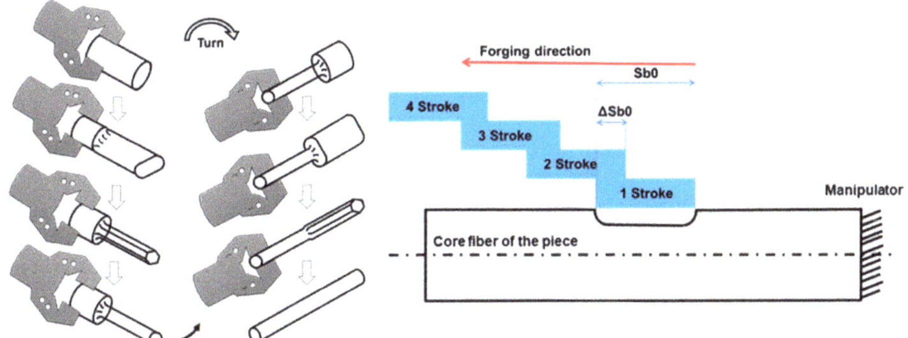


**Figure 1.** Schematic representation of the simulated forging process.

Simulation of the stretching forging process of a 42CrMo4 steel has been performed by an FE model developed using the commercial code MSC.Marc. Such code is well known for being characterized by a high accuracy tool for closed and open die forging process simulation [33–35]. The strokes start in the central part of the piece and proceed until the 4th stroke as shown in Figure 1.

In order to have accurate predictions from the model it is important to adopt proper material models to describe the flow stress. For this reason, laboratory compression tests are usually conducted at the Gleeble on cylindrical specimens with different temperatures, strains, strains rates, post deformation holding times to characterize the flow stress, static recrystallization and grain size evolution during forging. The rheology has been modelled following the model in [36].

At a first stage, some forging relevant parameters have been chosen in order to simulate the deformation process, such as: ingot diameter D, on contact die length $Sb_0$, forged piece temperature, percentage reduction. The chosen die has a flat shape, and the numerical values of the parameters adopted to implement the FEM model are shown in Table 1, which corresponds to about 600 simulations after Design of Experiment (DOE).

In Figure 2 it is possible to notice that the total equivalent plastic strain reaches the maximum value below the right side of the flat die. This effect is mainly due to the presence of the manipulator that, blocking the ingot as shown in Figure 1, causes a peak of plastic strain of about 1.02 (the amount of the deformation depends on forging conditions). This effect is in common for every simulated case.

**Table 1.** Parameters of FEM simulation for stretching forging process for round ingot.

| Forging Parameter | Minimum Value | Maximum Value | Step |
|---|---|---|---|
| Temperature [°C] | 800 | 1200 | 100 |
| $Sb_0$ [mm] | 150 | 750 | 150 |
| Reduction [%] | 5.0 | 25.0 | 2.5 |
| $\Delta Sb_0$ [%] (Pitch [%] respect $Sb_0$) | 10 (90%) | 50 (50%) | (20%) |
| Ingot initial diameter [mm] | 300 | 1500 | 300 |

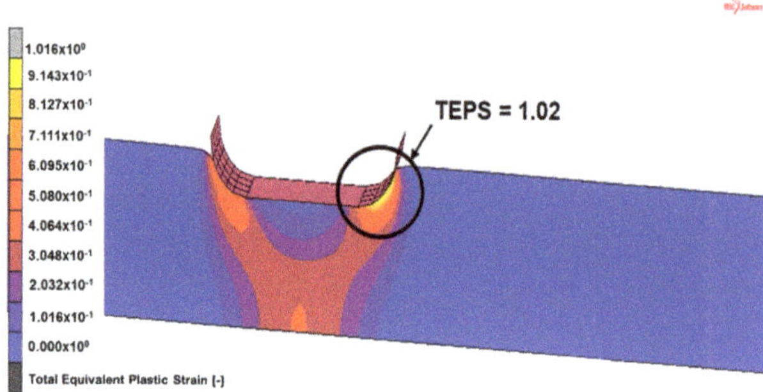

**Figure 2.** Effect of the presence of manipulator during forging process for a $Sb_0$ of 300 mm.

An example of FEM 3D map results in terms of total equivalent plastic strain at the core of the piece is reported in Figure 3. The model was implemented considering a double symmetry in longitudinal and radial directions and a constraint as in Figure 1 in order to consider the presence of the manipulator during the forging process. From a thermic point of view, the calculation was carried out considering isothermal condition, without exchange between ingot and external environment and tools in order to separate the thermal effect from the mechanical one on the plastic strain at the core of the ingot.

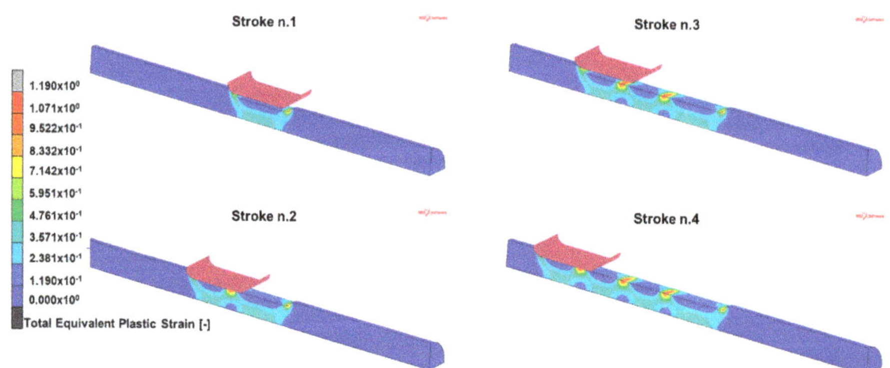

**Figure 3.** Total equivalent plastic strain at the core of the piece for an ingot with a diameter of 300 mm, 25% reduction, for Sb0 of 300 mm and a $\Delta$Sb0 of 10% at T = 1200 °C for the first series of strokes.

Figures 2 and 3 show the typical distribution of plastic deformation through the ingot radius and longitudinal direction with the maximum value of deformation reached in the contact area between tools and ingot closest to the manipulator. In the thickness, the deformation distribution takes on the classic v-shape with the relative maximum on the core fiber [37] for the considered ratio between contact die length and ingot diameters ($Sb_0$/D).

## 2.2. Analytical Model for Open Die Process Simulation

A prior analysis of FEM results in terms of plastic strain at the core of the forged piece (Figure 4) has been carried out in order to represent the plastic strain at the center of the piece.

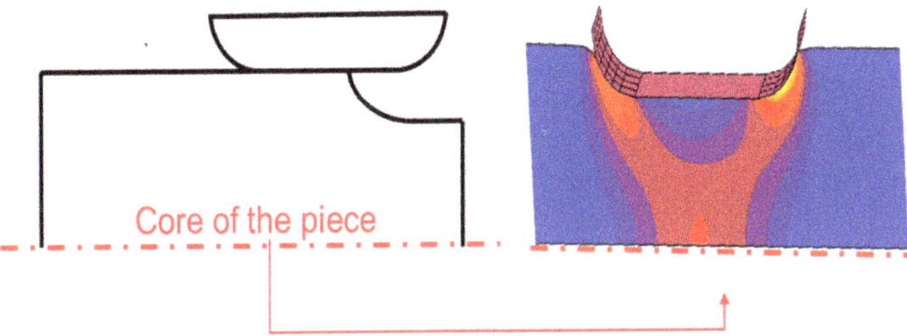


**Figure 4.** Schematic representation of forged piece simulated; results have been token at the core of the piece in terms of plastic strain as a function of the length.

An example of output of the FEM thermo-mechanical simulation follows the evolution reported in Figure 5 concerning the evolution of plastic strain as a function of length at 1200 °C, $Sb_0$ = 300 mm, reduction = 25%. As shown in Figure 5, the simulated forging process is characterized by a first stroke characterized by a major contact zone due to the fact that during the first stroke the material is not yet deformed. The last three strokes follow the same evolution.

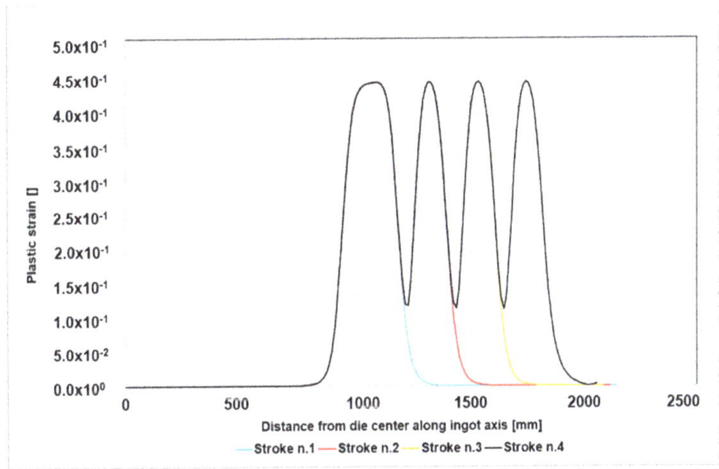


**Figure 5.** Evolution of plastic strain as a function of length at 1200 °C, $Sb_0$ = 300 mm, reduction = 25%, ingot diameter equal to 300 mm, pitch 90%.

In this work a separation of the strokes was performed in order to distinguish each one and apply the analytical model and the neural network separately. Moreover, because the strokes following the first follow similar behavior in terms of maximum plastic strain, only the first and the second strokes could be considered in the analysis.

According to [37] the distance between the growth and decay phase of plastic strain on ingot core fiber increases while the ratio Sb0/D decreases. This effect is showed in Figure 5: comparing the first stroke (Sb0/D = 1.0) and the subsequent (Sb0/D = 0.9), it is possible to note that, with a pitch equal to

90%, the $Sb_0$ is reduced to 270 mm. A double sigmoidal function has been chosen to represent the evolution of the core fiber plastic strain of a single stroke [38,39]. The sum of two hyperbolic tangents (Equation (1)) has been implemented in order to better reproduce the plastic strain evolution at the core of the forged piece along the ingot axis. The growth phase and decay phase of plastic strain are represented by Equations (2) and (3). Equation (1) varies from 0 to 2 and is continuously derivable and defined throughout the Real numbers domain, therefore it can be used without problems in an optimization system.

$$\varepsilon^p_{tot} = M * \left( \tan h\left(\frac{x - C_1}{D_1}\right) + \tan h\left(-\frac{x - C_2}{D_2}\right) \right) \tag{1}$$

$$\varepsilon^p_1 = \tan h\left(\frac{x - C_1}{D_1}\right) \tag{2}$$

$$\varepsilon^p_2 = \tan h\left(-\frac{x - C_2}{D_2}\right) \tag{3}$$

Coefficients in Equations (1)–(3) are respectively:

- $C_1$ and $C_2$ represent the middle points of growth and decay phase respectively. In order to identify the inflection point for each sigmoidal branch, an analysis has been carried out on FEM results, identifying the point with the 50% of the maximum plastic strain at the core fiber.
- $D_1$ and $D_2$ represent the slopes of the growth and decay branches of the function.
- M is a multiplier coefficient. The Equation (1) varies in a range between 0 and 2, thus the coefficients M brings the maximum of double-sigmoidal curve to the maximum of plastic strain.

The coefficients $C_1$, $C_2$, $D_1$, $D_2$ and M have been obtained by fitting of double-sigmoidal model on the FEM results in terms of total equivalent plastic strain along the core fiber for each forging configuration considered in this analysis. The obtained coefficients do not have an identified mathematical dependence by forging parameters. The fitting of mathematical model described by Equation (1) to the individual conditions of the forging process could be carried out using a fragmented, look-up table-based approach. This approach has disadvantages due to the required size for the look-up tables and the lack of interpolation capability. The application of neural networks within the control strategy, setup model, and optimization tool has significantly reduced such kind of problems [40]. The obtained coefficients were used to train the proposed neural network.

*2.3. Forecasting Models Based on Artificial Neural Networks (ANNs)*

The artificial neural networks (ANNs) can be considered nonlinear regressive models, realizing the correlation between a set of independent variables and a set of dependent variables.

Neural networks are mathematical models able to learn from empirical data collected in some problem domain by approximating sample of it in a data set, without any assumption about the physical laws, systems inspired by biological neural networks that constitute brains of animals. This correlation between variables is achieved through a training process during which a data set containing both independent and dependent variables is used to iteratively adapt the internal structure of the neural model to its purpose [41].

Because the functional relationship between causes and effect of the physical phenomena is extracted from the samples, the performance provided by the ANN model is related to the range of variability and accuracy of the data set. Such models are completely different from a Deterministic Model that requires a priori knowledge of the relationships of a system, based on First Principles typically derived from physical, chemical or biological principles. For this reason, high complexity of a problem or the unsatisfactory performance of other techniques (e.g., deterministic model, Linear and Not-Linear multiple regression) are condition for a suitable application of an ANN model. For this reason, artificial neural network models are considered a sort of Black Boxes which do not account the mathematical expression describing the physical phenomenon once the training procedure is finished.

The ANN performances are comparable to the correspondent Deterministic model, and are chosen for a fast development of models due to the flexibility to codify relationships between any set of physical variables (real, discrete and Boolean), and an easy integration with resident control system and SW. The Requirement for a stochastic ANN model is mainly the Input/Output (I/O) functional relationship between causes and effect and a set of process observation covering a wide enough range of variability of the variables [31].

Artificial neural networks can be collected in families according to the learning and recall algorithms. The network adopted in this work belongs to the multi-layer perceptron (MLP) family whose learning algorithm is back propagation (BP) [42]. The back-propagation learning algorithm provides the optimal configuration of the weights by calculating the error between the target and the network response. The root mean squared (RMS) error has been adopted as index of performance both for each single output variable and for the output as a whole.

Since these mathematical models are based on data observation, their performance is strongly conditioned by the quality of the data itself. Because the data are provided by the calculation of a FE model and simulations are defined by means of a DOE, no data duplication (similarities) or scattering are present in the data. In this condition, no correction (elimination of similarities) or filtering action are required [41,42]. Clustering analysis was carried out in order to define and identify the cluster, and then for the definition of best neural network topology and number of neural networks. The independent variables of ANN are the forging parameters in terms of workpiece temperature and diameter, $Sb_0$, fitch, reduction and stroke number while the output is in terms of the coefficients of double-sigmoidal function that models total equivalent plastic strain. Input and output data have been normalized within the range 0 and 1 with a linear function between the minimum and maximum value of each quantities, Table 2.

Each node in ANN is fully connected to the nodes of the following layer (hidden or not) through a sigmoidal transferring function and weights whose value is adapted during the learning phase to encode on them the knowledge of the forging process described by the used dataset [42].

The implemented ANN is composed of 1 bias node and a single hidden layer characterized by 13 hidden nodes.

The initial data, that consisted in about 600 examples, has been divided into three groups:

- Training: about 400 examples;
- Validation: about 150 examples;
- Test: about 50 examples.

The examples have been subdivided considering the three main clusters identified during the data analysis.

**Table 2.** Normalized values for artificial neural network.

| Variable | Min Value | Max Value | Min Normalized Value | Max Normalized Value |
|---|---|---|---|---|
| $Sb_0$ | 75 | 750 | 0.1 | 0.9 |
| Temperature | 800 | 1200 | 0.1 | 0.9 |
| Reduction [%] | 5 | 25 | 0.1 | 0.9 |
| $C_1$ | −384 | −28 | 0.1 | 0.9 |
| $D_1$ | 19 | 60 | 0.1 | 0.9 |
| $C_2$ | 9.5 | 350 | 0.1 | 0.9 |
| $D_2$ | 27 | 60 | 0.1 | 0.9 |
| M | 0 | 0.24 | 0.1 | 0.9 |

## 3. Results and Discussion

Figures 6–8 report the comparison between the coefficient of Equation (1) fitted on FE results and calculated by ANN concerning the 50 examples used for the tests. The figures show that the data groups into the three identified cluster. The scatter plot related to the C1 and C2 coefficients used to train the ANN coefficients, and the coefficients trained by ANN is reported in Figure 6a,b. Furthermore, $R^2$ is reported in such figures. Results are characterized by little dispersion and the good agreement between the coefficient forecasted by ANN and fitting on FE results is confirmed. In fact, concerning C1, the $R^2$ is approximately 1. Looking at Figures 7 and 8, it is possible to notice that the $R^2$ always remains high, never falling below 0.997.

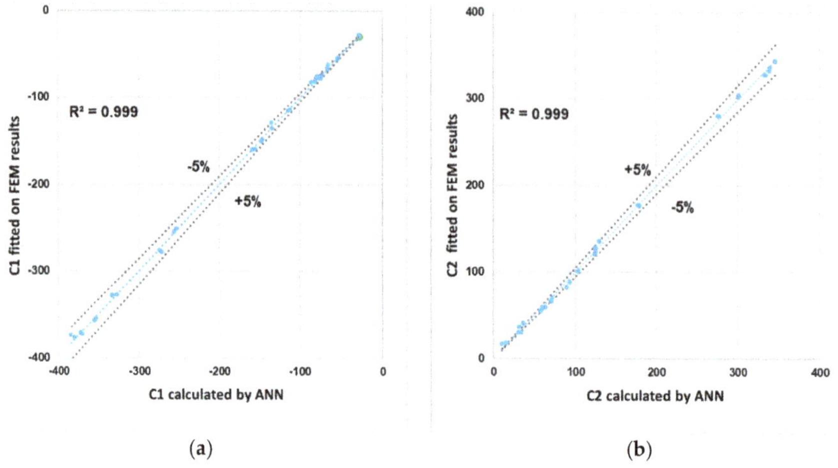


**Figure 6.** Plot of C1 from the analytical model C1 (target) versus the ANN trained C1 in blue with error bands at +5% and −5% in grey (**a**) and the analytical model C2 (target) versus the ANN trained C2 in blue with error bands at +5% and −5% in grey (**b**).

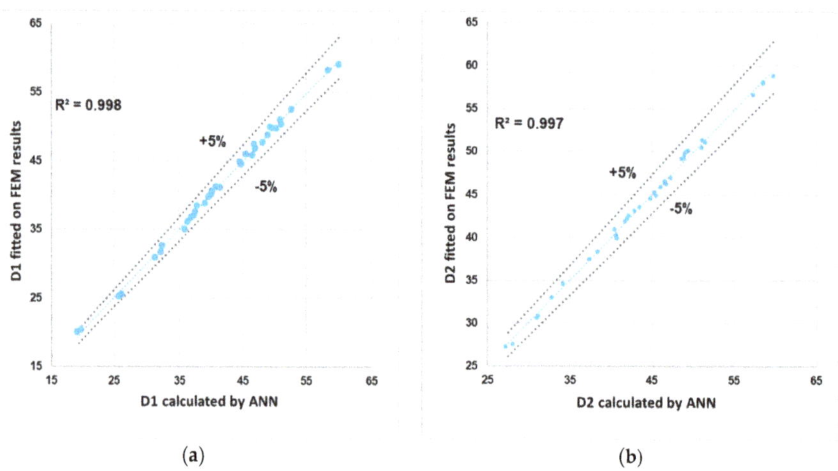


**Figure 7.** Plot of D1 from the analytical model D1 (target) versus the ANN trained D1 in blue with error bands at +5% and −5% in grey (**a**) and the analytical model D2 (target) versus the ANN trained D2 in blue with error bands at +5% and −5% in grey (**b**).

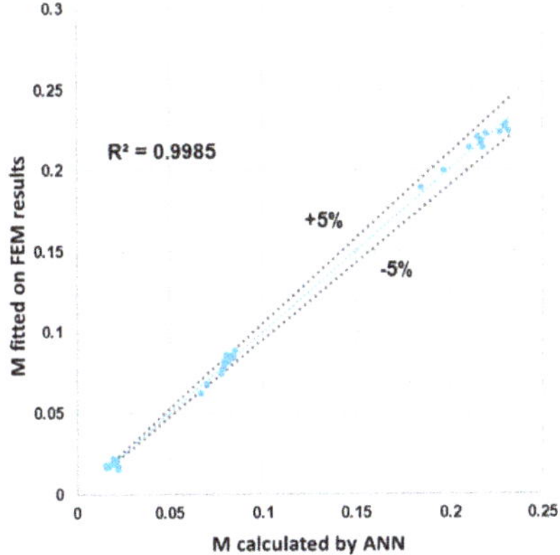

**Figure 8.** Plot of M from the analytical model M (target) versus the ANN trained M in blue with error bands at +5% and −5% in grey.

In the following figures there is a comparison between FEM (black line), analytical model (as obtained by fitting of FEM data, red dot line) and analytical model with coeffects predicted by ANN (blue line) is reported in Figures 9–14 in terms of plastic strain dependence on arch length. The RMS deviation (green line) between the neural network and the analytical model results is also reported.

The above comparison for the case corresponding to initial ingot diameter equal to 300 mm, $Sb_0$ = 150 mm, reduction = 5% at 800 °C and 1200 °C is reported in Figure 9a,b, respectively. The proposed modeling approach provides the possibility to have two different slopes of the growth and decay phase of plastic strain (e.g., different $D_1$ and $D_2$ coefficients). This allows to consider the effect of the manipulator nevertheless it is possible to notice that the error peak is in correspondence of the presence of the manipulator, therefore in the decay branch of the double-sigmoid model. This is expected considering that the manipulator induces a small variation from the chosen sigmoidal shape, shifting the inflection point towards 60–70% of the maximum strain value. However, the hybrid approach prosed ANN plus analytic model is able to reproduce the FEM curve with a good accuracy both temperature, 800 °C and 1200 °C (maximum RMS value = 0.07).

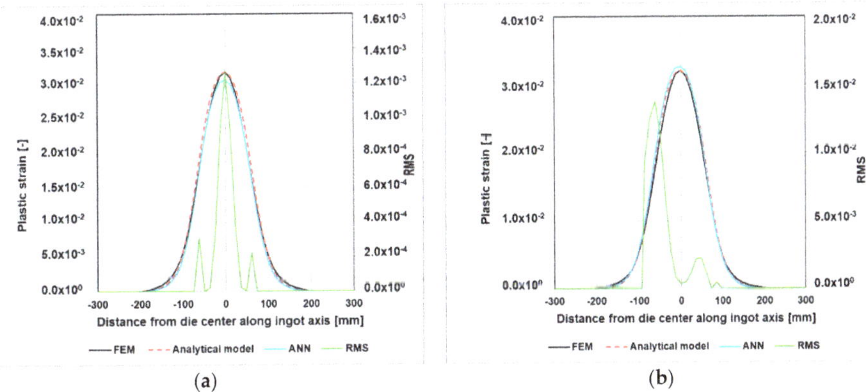

**Figure 9.** Total equivalent plastic strain as a function of length for $Sb_0$ = 150 mm, reduction of 5%, first stroke at (**a**) 800 °C, (**b**) 1200 °C.

A similar trend is shown in Figure 10, concerning the same initial ingot diameter and $Sb_0$ = 150 mm, but with higher reductions (equal to 25%). Also, in this case 800 °C and 1200 °C temperatures are compared. Figures 9 and 10 show that for $Sb_0$ equal to 150 mm temperature variation little affect the maximum of plastic strain and shape of plastic strain on core fiber. Differences on maximum value of plastic strain are equal to 1.4% and 4.4% for the reduction equal to 5% and 25%, respectively, with a material softening from 800 °C to 1200 °C equal to 70% at strain = 0.3 and strain rate = 1 s$^{-1}$ from 205 MPa to 62 MPa respectively, Figure 11 [43]. his so low influence of rheological behavior is certainly due to of isothermal FE modeling of forging but leads to the conclusion that the proposed approach can be considered independent of the material in the first approximation.

The same is shown in Figures 12 and 13, where plastic strain for first stroke of forging of ingot with 300 mm of initial diameter of the higher reduction considered in this analysis is compared for temperature of 800 °C and 1200 °C considering and $Sb_0$ = 300 mm and 750 mm, respectively. Increasing $Sb_0$ the maximum plastic deformation value on core fiber increases. Moreover, a strong shift away from the growth and decay branches of plastic strain is observed increasing the $Sb_0/D$ ratio, as described before. Also, in this case of even larger size of the die, the ANN approach appears to be in good agreement with FEM results, with minimal error.

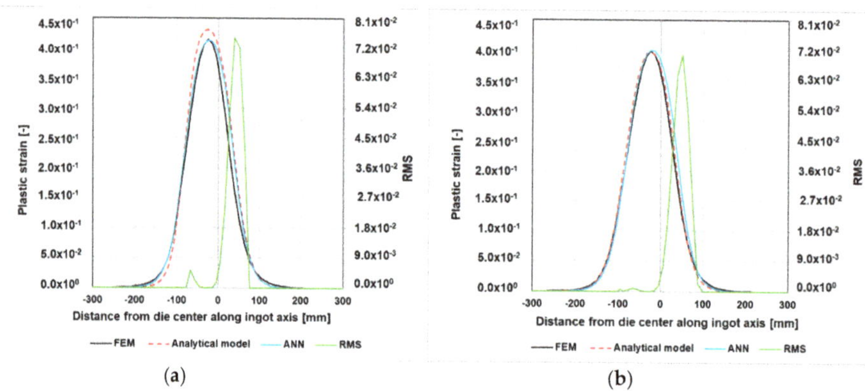

(a)        (b)

**Figure 10.** Total equivalent plastic strain as a function of length for $Sb_0$ = 150 mm, reduction of 25%, first stroke at (**a**) 800 °C, (**b**) 1200 °C.

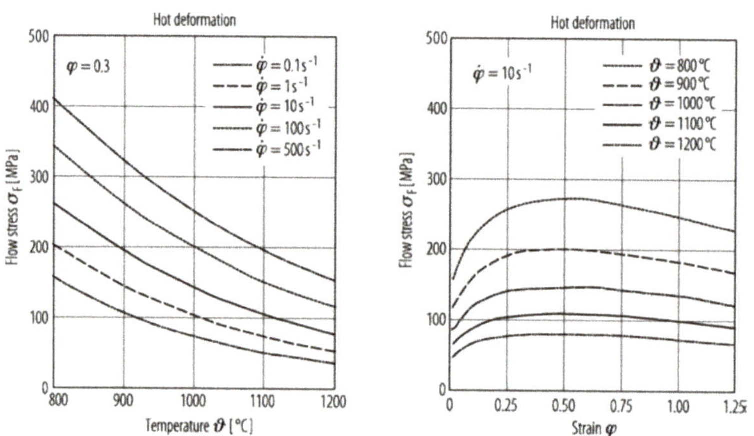

**Figure 11.** 42CrMo4. Flow stress curves by hot forming. Figure 5.309 in [44].

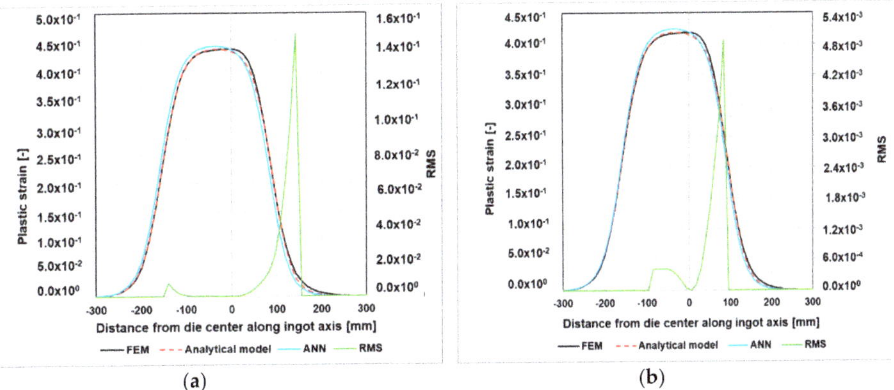

**Figure 12.** Total equivalent plastic strain as a function of length for $Sb_0 = 300$ mm, reduction of 25% for first stroke at (**a**) 800 °C, (**b**) 1200 °C.

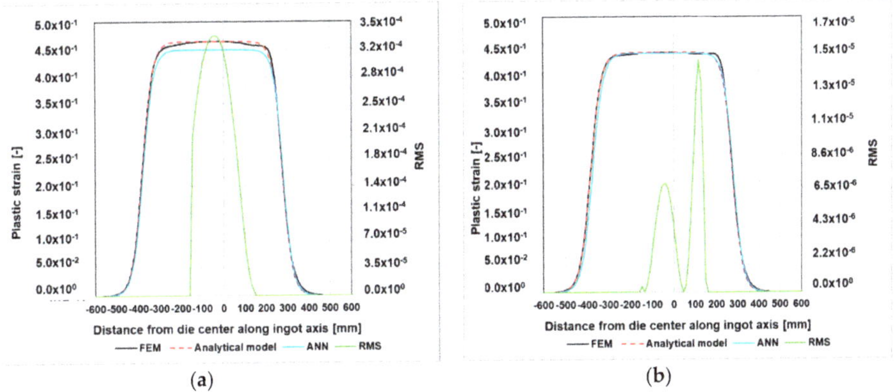

**Figure 13.** Total equivalent plastic strain as a function of length for $Sb_0 = 750$ mm, reduction of 25% for first stroke at (**a**) 800 °C, (**b**) 1200 °C.

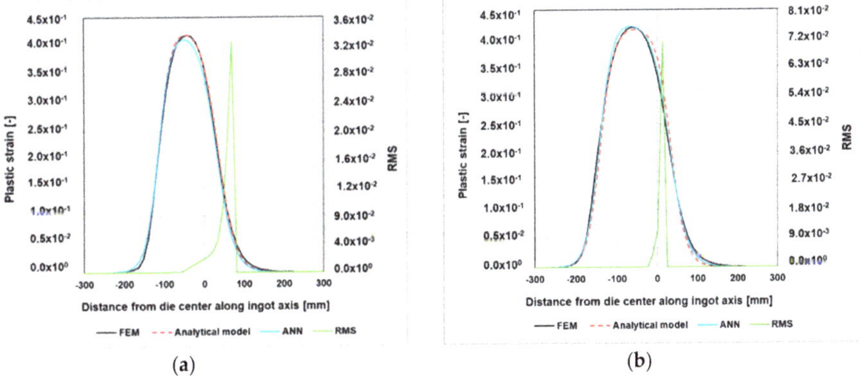

**Figure 14.** Total equivalent plastic strain as a function of length for $Sb_0 = 300$ mm, reduction of 25% at 1200 °C for second (**a**) and third stroke (**b**).

For the stroke subsequent the first the fitch, and the resulting overlapping of strokes, affects directly the actual $Sb_0$ reducing it. The comparison in Figure 14 shows as the stroke subsequent to the first could be considered, in first approximation, equal to a first stroke with a lower value of $Sb_0$. However, the developed model has been differentiated for the first stroke and the subsequent ones in order to take into account that in the stroke subsequent to the first there are parts of die initially not in contact with the ingot. These will touch parts of ingot already deformed or partially deformed, e.g., deformed by die fillet radius in the previous stroke, upsetting it. This results in an improvement in the performance of the model, e.g., a better agreement between the FEM results and the analytical model with coefficients calculated with the ANN.

## 4. Conclusions

In this paper a hybrid approach is proposed, able to describe the plastic strain behavior at the core fiber of an open die forged round shape component. Such a method takes into account the following parameters: ingot diameter, die length $Sb_0$, reduction for each stroke and forging temperature. The aim of this approach is to provide a rapid tool faster than the commonly used FEM method but with the same accuracy class, making therefore it suitable for the rapid design of online forging processes. This was accomplished by means of a thermo-mechanical FEM model implemented in the MSC.Marc commercial code in order to simulate the open die forging process.

Starting from the results obtained through FEM simulations, a set of equations describing the plastic strain at the core fiber of the piece have been identified depending on forging parameters. An artificial neural network has been trained to provide the double-sigmoid coefficients as function of forging parameters. The maximum error by proposed model prediction is found at the peak deformation and on the decay branch due to the presence of the manipulator. The results analysis showed the low dependence of strain on core fiber on the material rheology. In the first approximation, therefore, the material properties could be neglected, but this becomes fundamental when microstructural and metallurgical effects are also considered in the optimization model of forging.

The described approach proposes therefore a rapid method aimed to design and optimize a forging open die process, favoring its adoption in industrial applications.

**Author Contributions:** Conceptualization, S.M. and L.L.; methodology, S.M. and L.L.; software, S.M. and L.L.; validation, S.M. and L.L.; formal analysis, S.M. and R.M.; investigation, S.M. and L.L.; resources, L.L.; data curation, S.M. and R.M; writing—original draft preparation, S.M.; writing—review and editing, S.M., A.D.S., L.L., G.Z. and R.M.; visualization, A.D.S., L.L.; supervision, A.D.S. All authors have read and agreed to the published version of the manuscript.

**Funding:** This research received no external funding.



## References


1. Di Schino, A. Manufacturing and application of stainless steels. *Metals* **2020**, *10*, 327. [CrossRef]
2. Pezzato, L.; Gennari, C.; Chukin, D.; Toldo, M.; Sella, F.; Toniolo, M.; Zambon, A.; Brunelli, K.; Dabalà, M. Study of the effect of multiple tempering in the impact toughness of forged S690 structural steel. *Metals* **2020**, *10*, 507. [CrossRef]
3. Di Schino, A.; Di Nunzio, P.E.; Lopez Turconi, G. Microstructure Evolution during Tempering of Martensite in a Medium-C Steel. *Mater. Sci. Forum* **2007**, *558*, 1435–1441. [CrossRef]
4. Di Schino, A. Analysis of phase transformation in high strength low alloyed steels. *Metalurgija* **2017**, *56*, 349–352.
5. Di Schino, A.; Alleva, L.; Guagnelli, M. Microstructure evolution during quenching and tempering of martensite in a medium C steel. *Mater. Sci. Forum* **2012**, *715–716*, 860–865. [CrossRef]
6. Di Schino, A.; Kenny, J.M.; Abbruzzese, G. Analysis pf the recrystallization and grain growth processes in AISI 316 stainless steel. *J. Mat. Sci.* **2002**, *37*, 5291–5298. [CrossRef]

7.  Rufini, R.; Di Pietro, O.; Di Schino, A. Predictive simulation of plastic processing of welded stainless steel pipes. *Metals* **2018**, *8*, 519. [CrossRef]

8.  Di Schino, A.; Testani, C. Corrosion behavior and mechanical properties of AISI 316 stainless steel clad Q235 plate. *Metals* **2020**, *10*, 552. [CrossRef]

9.  Gloria, A.; Montanari, R.; Richetta, M.; Varone, A. Alloys for Aeronautic Applications: State of the Art and Perspectives. *Metals* **2019**, *9*, 662. [CrossRef]

10. Di Schino, A.; Gaggiotti, M.; Testani, C. Heat treatment effect on microstructure evolution in 7% Cr steel for forging. *Metals* **2020**, *10*, 808. [CrossRef]

11. Schafrik, R.E.; Walsson, S. Challenges for high temperature materials in the new millennium. In Proceedings of the Eleventh International Symposium on Superalloys, Champion, PA, USA, 14–18 September 2008; p. 3.

12. Di Schino, A.; Valentini, L.; Kenny, J.M.; Gerbig, Y.; Ahmed, I.; Haefke, H. Wear resistance of a high-nitrogen austenitic stainless steel coated with nitrogenated amorphous carbon films. *Surf. Coat. Technol.* **2002**, *161*, 224–231. [CrossRef]

13. Dindorf, R.; Wos, P. Energy-Saving Hot Open Die Forging Process of Heavy Steel Forgings on an Industrial Hydraulic Forging Press. *Energies* **2020**, *13*, 1620. [CrossRef]

14. Technovio. *Global Forging Market 2018–2022*; Technovio: Toronto, ON, Canada, 2019.

15. Zitelli, C.; Folgarait, P.; Di Schino, A. Laser powder bed fusion of stainless-steel grades: A review. *Metals* **2019**, *9*, 731. [CrossRef]

16. Mancini, S.; Langellotto, L.; Di Nunzio, P.E.; Zitelli, C.; Di Schino, A. Defect reduction and quality optimisation by modelling plastic deformation and metallurgical evolution in ferritic stainless steels. *Metals* **2020**, *10*, 186. [CrossRef]

17. Qiu, Y.; Park, S.C.; Cho, H.Y. Prediction of forming limits in cold open-die extrusion process. *Trans. Korean Soc. Mech. Eng. A* **2020**, *44*, 435–441. [CrossRef]

18. Harris, N.; Shahriari, D.; Jahazi, M. Analysis of Void Closure during Open Die Forging Process of Large Size Steel Ingots. *Key Eng. Mater.* **2016**, *716*, 579–585. [CrossRef]

19. Di Schino, A.; Di Nunzio, P.E. Metallurgical aspects related to contact fatigue phenomena in steels for back up rolling. *Acta Metall. Slovaca* **2017**, *23*, 62–71. [CrossRef]

20. Sharma, D.K.; Filipponi, M.; Di Schino, A.; Rossi, F.; Castaldi, J. Corrosion behavior of high temperature fuel cells: Issues for materials selection. *Metalurgija* **2019**, *58*, 347–351.

21. Wolfgarten, M.; Rosenstock, D.; Schaeffer, L.; Hirt, G. Implementation of an open-die forging process for large hollow shafts for wind power plants with respect to an optimized microstructure. *AIM* **2015**, *107*, 43–49.

22. Choi, S.K.; Chun, M.S.; Van Tyne, C.J.; Moon, Y.H. Optimization of open die forging of round shapes using FEM analysis. *J. Mater. Process. Technol.* **2006**, *172*, 88–95. [CrossRef]

23. Obiko, J.; Mwema, F.M. Stress and Strain Distribution in the upsetting process. In *Handbook of Research on Advancements in Manufacturing, Materials, and Mechanical Engineering*; IGI Global: Hershey, PA, USA, 2020; pp. 288–301.

24. Rosenstoc, D.; Recker, D.; Hirt, G.; Steingießer, K.; Rech, R.; Gehrmann, B.; Lamm, R. Application of a Fast Calculation Model for the Process Monitoring of Open Die Forging Processes. *Key Eng. Mater.* **2013**, *554*, 248–263. [CrossRef]

25. Siemer, E.; Nieschwitz, P.; Kopp, R. Quality-optimized process control in open-die forging. *Stahl Eisen* **1986**, *106*, 383–387.

26. Napoli, G.; Di Schino, A.; Paura, M.; Vela, T. Colouring titanium alloys by anodic oxidation. *Metalurgija* **2018**, *57*, 111–113.

27. Kim, P.H.; Chun, M.S.; Yi, J.J.; Moon, Y.H. Pass schedule algorithms for hot open die forging. *J. Mater. Process. Technol.* **2002**, *130*, 516–523. [CrossRef]

28. Jarl, M. FEM simulation of drawing out in open die forging. *Steel Res. Int.* **2004**, *75*, 812–817.

29. Recker, D.; Franzke, M.; Hirt, G. Fast models for online optimization during open die forging. *CIRP Ann. Manuf. Technol.* **2011**, *60*, 295–298. [CrossRef]

30. Franzke, M.; Recker, D.; Hirt, G. Development of a Process Model for Online optimization of Open Die Forging of Large Workpieces. *Steel Res. Int.* **2008**, *79*, 753–757. [CrossRef]

31. Haykin, S. *Neural Networks: A Comprehensive Foundation*, 2nd ed.; McMaster University: Hamilton, ON, Canada, 1994.

32. Russell, S.; Norvig, P. *Artificial Intelligence: A Modern Approach*, 2nd ed.; Pearson Education: Upper Saddle River, NJ, USA, 2003.

33. Hung, C.; Kobayashi, S. Three-dimensional finite element analysis on open-die block forging design. *J. Eng. Ind.* **1992**, *114*, 459–464. [CrossRef]

34. Skunca, M.; Skakun, P.; Keran, Z.; Sidjanin, L.; Math, M.D. Relations between numerical simulation and experiment in closed die forging of a gear. *J. Mater. Process. Technol.* **2006**, *177*, 256–260. [CrossRef]

35. Zhang, Z.; Xie, J. A numerical simulation of super-plastic die forging process for Zr-based bulk metallic glass spur gear. *Mater. Sci. Eng. A* **2006**, *433*, 323–328. [CrossRef]

36. Hensel, A.; Spittel, T. *Kraft und Hitsbedarf Bildsamer Formgebungsverfahren*; VEB Deutscher Verlag für Grundstoffindustrie: Leipzig, Germany, 1978.

37. Shah, K.N.; Kiefer, B.V.; Gavigan, J.J. Finite Element Simulation of Internal Void Closure in Open-Die Press Forging. *Mater. Manuf. Process* **1986**, *1*, 501–516. [CrossRef]

38. Matsumoto, Y.P.; Chiba, A. Correcting the stress strain curve in the stroke-rate controlling forging process. *Metall. Mater. Trans. A* **2009**, *40*, 1203–1209.

39. Caglar, M.U.; Teufel, A.I.; Wilke, C.O. Sicegar: R package for sigmoidal and double-sigmoidal curve fitting. *PeerJ* **2018**, *6*, e4251. [CrossRef] [PubMed]

40. Schlang, M.; Feldkeller, B.; Lang, B.; Poppe, T.; Runkler, T. Neural computation in steel industry. In Proceedings of the 1999 European Control Conference (ECC), Karlsruhe, Germany, 31 August–3 September 1999; pp. 2922–2927.

41. Bishop, C.M. *Neural Networks for Pattern Recognition*, 2nd ed.; Oxford University Press: New York, NY, USA, 1995.

42. Wasserman, P.D. *Advanced Methods in Neural Computing*; John Wiley & Sons Inc.: New York, NY, USA, 1993.

43. Mehta, S.B.; Gohil, D.B. Computer Simulation of Forging Using the Slab Method Analysis. *Int. J. Sci. Eng. Res.* **2011**, *2*, 1–5.

44. Spittel, M.; Spittel, T. *Landolt-Börnstein–Numerical Data and Functional Relationships in Science and Technology, Group VIII: Advanced Materials and Technologies Volume 2, Materials–Subvolume C Metal Forming Data, Ferrous Alloys*; Springer: Berlin/Heidelberg, Germany, 2009.

**Publisher's Note:** MDPI stays neutral with regard to jurisdictional claims in published maps and institutional affiliations.

*Article*

# Study of the Effect of Multiple Tempering on the Impact Toughness of Forged S690 Structural Steel

Luca Pezzato [1], Claudio Gennari [1], Dmitry Chukin [2], Michele Toldo [3], Federico Sella [3], Mario Toniolo [3], Andrea Zambon [4], Katya Brunelli [1,5] and Manuele Dabalà [1,*]

[1] Department of Industrial Engineering, University of Padua, Via Marzolo 9, 35131 Padova, Italy; luca.pezzato@unipd.it (L.P.); claudio.gennari@studenti.unipd.it (C.G.); katya.brunelli@unipd.it (K.B.)
[2] Nosov Magnitogorsk State Technical University, pr. Lenina, 38, 455000 Magnitogorsk, Russia; chukindmitry@gmail.com
[3] FOC Ciscato S.p.a., Via Pasin 1, 36010 Velo D'astico (VI), Italy; m.toldo@foc.it (M.T.); f.sella@foc.it (F.S.); m.toniolo@foc.it (M.T.)
[4] Department of Management and Engineering, University of Padova, Stradella San Nicola 3, 36100 Vicenza, Italy; a.zambon@unipd.it
[5] Department of Civil, Environmental and Architectural Engineering, University of Padova, Via Marzolo 9, 35131 Padua, Italy
* Correspondence: manuele.dabala@unipd.it; Tel.: +39-0498-275-749

Received: 13 March 2020; Accepted: 10 April 2020; Published: 14 April 2020

**Abstract:** During the production of forged metal components, the sequence of heat treatments that are carried out, as well as hot working, remarkably influences mechanical properties of the product, in particular impact toughness. It is possible to tailor impact toughness by varying tempering temperature and soaking time after hardening treatment, widening the application range of structural steels. In this work, we consider the effects of a second tempering treatment on the microstructural properties and impact toughness of a structural steel EN 10025-6 S690 (DIN StE690, W. n: 1.8931). The steel was first forged and quenched in water after austenitization at 890 °C for 4 h. After quenching different tempering treatments were performed, at 590 °C in single or multiple steps. The effect of these treatments was evaluated both in microstructural terms, by means of optical microscopy, scanning and transmission electron microscopy and X-ray diffraction, and in terms of impact toughness. The mechanical behavior was correlated with the microstructure and a remarkable increase in impact toughness was found after the second tempering treatment due to carbide shape change.

**Keywords:** tempering; impact toughness; carbides; heat treatment; bainite; steel

---

## 1. Introduction

S690 is a low-alloy high strength steel mainly employed in structural applications. One of the typical uses of this steel grade is offshore applications, where high impact toughness even at low temperatures is required. In these applications, the steel is supplied with a mainly bainitic microstructure. Moreover, considering the low carbon content this steel is also characterized by a relatively good weldability [1,2]. In more detail, in order to obtain the desired properties, high strength steels are formed by forging and thereafter undergo a quenching and tempering treatment [3–5]. One of the key mechanical properties of steels employed in structural applications is impact toughness. However, impact toughness is not related only to the chemical composition of the metal but can be significantly modified by deformation route or/and heat treatments [6]. Among the different heat treatments, quenching and tempering are used extensively in the forging industry. Tempering efficiency in forging is important both for mechanical properties achievement as well as for environmental and economic reasons. Since treated parts and equipment are commonly huge, the related time for the treatment and natural gas consumption is high and onerous, hence any optimization in the treatment cycle (and of course on furnaces insulation) can reduce atmospheric emissions and produce economic saving. The tempering treatments performed for long soaking times seem to significantly increase the impact toughness of different steel grades [7,8]. This is due to bainitic carbides spheroidization, that produces an increase in mechanical properties by reducing stress concentration [9,10]. The modification of carbides morphology in fully bainitic steels depends on the carbides type. For example, as evidenced by Depinoy et al. [11], for a 2.25Cr-1Mo steel four types of carbides can precipitate depending on tempering temperature and time: $M_3C$, $M_2C$, $M_7C_3$ and $M_{23}C_6$. $M_3C$ is the least stable carbide and is characterized by lenticular-globular morphology, $M_2C$ and $M_7C_3$ are characterized by a needle like and rhombus shaped morphology, $M_{23}C_6$ is the equilibrium carbide and possesses a rod-like morphology. Generally, a more spherical shape of the carbides can lead to an increase in the impact toughness [12].

Impact toughness of quenched and tempered components can be strongly decreased also by the presence of retained austenite after tempering due to: (i) precipitation of interlath cementite aided by partial thermal decomposition of interlath films of retained austenite; (ii) subsequent deformation-induced transformation on loading of remaining interlath austenite, which has become mechanically unstable due to carbon depletion as a consequence of carbides precipitation as demonstrated by Horn et al. [13]. This problem can also affect S690 steel, as observed by Yen et al. [14] who proposed a new steel composition for offshore applications to overcome the problem. Another possible approach that can solve the problem is to decompose retained austenite using multiple tempering treatment [15,16].

However, even if S690 steel is a commonly used structural steel, very few works regarding the effect of heat treatments on its impact toughness and the microstructure can be found in literature. In particular, one work by Duan et al. regarding the effect of quenching temperature on the microstructure [17], one by Dong et al. [18] about the effect of the quenching and partitioning treatment, and one by Quin et al. [19] where different heat treatments, with an orthogonal experiment, have been performed, were found in literature.

The aim of the present work is to study the difference between a single long-lasting tempering treatment compared to multiple tempering treatments on forged and quenched S690 steel and to verify whether a connection exists between the microstructural changes and the variations, aiming to an improvement in the impact toughness.

## 2. Materials and Methods

Test samples came from a production forging Bush (forging ratio of >4:1 from starting ingot), with dimensions $\Phi$2520 mm $\times$ $\Phi$2130 mm $\times$ H1417 mm, and weighing about 16 tons. A test ring was cut after heat treatment of the forging bush (austenitization at 890 °C for 4 h, quenching in water, cooling rate approximately 115 °C/s and tempering at 620 °C for 6 h) as reported in Figure 1A, and cut again to obtain samples 60 mm $\times$ 40 mm $\times$ 50 mm (Figure 1B) to perform heat treatments in lab scale. Standard V-notch impact test samples were machined transversally to the major forging direction, as reported in Figure 1C.

Composition of the steel is reported in Table 1. Considering the different heat treatments performed on the laboratory scale samples three parts of the test ring were austenitized at 890 °C for 4 h and quenched in agitated water at 30 °C. Two parts of the test ring were tempered at 590 °C for 6 h or at 590 °C for 10 h. Another part of the test ring was tempered at 590 °C for 6 h and tempered for a second time at 590 °C for other 4 h. In this way, the behaviors of samples with the same total tempering time but performed either in a unique treatment of 10 h or in two treatments of 6 h + 4 h were compared.

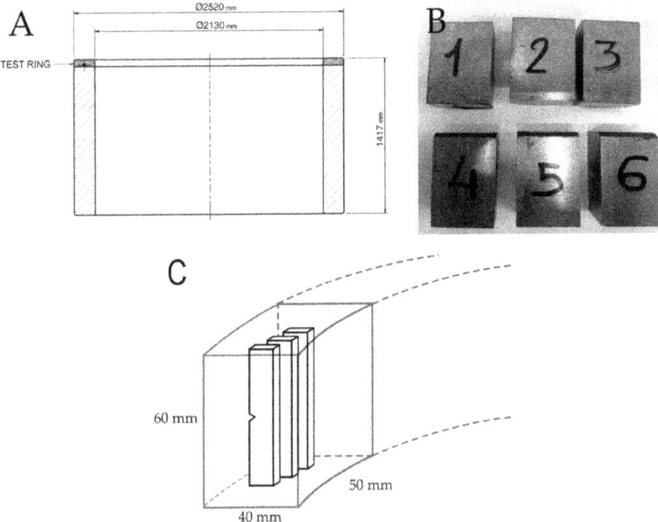

**Figure 1.** Drawing of the forging bush and of the position of the test ring (**A**); test ring sections used for the laboratory scale heat treatments (**B**); sketch of the extraction of the Charpy specimens from the test ring (**C**).

**Table 1.** Composition of the S690 steel (wt.%).

| C% | Mn% | Si% | Cr% | Ni% | Mo% | V% |
|------|------|------|------|------|------|-------|
| 0.14 | 1.29 | 0.31 | 1.11 | 1.11 | 0.27 | 0.044 |

The sequence of heat treatments performed in the case of the double tempering of laboratory scale test ring sections is summarized in Figure 2.

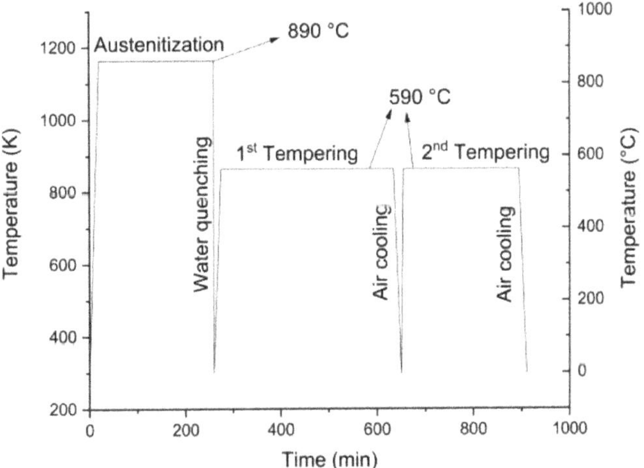


**Figure 2.** Heat treatment performed on the double tempered sample: austenitization at 890 °C for 4 h, first tempering at 590 °C for 6 h and second tempering at 590 °C for 4 h.

In the case of quenching and single tempering, carbides precipitation sequence was modeled with JMatPro® software (7.0.0, Sente Software Ltd., Guildford, UK).

Mechanical properties were evaluated by means of Vickers $HV_{0.2}$ micro-hardness tests using a Leitz Miniload 2 (Leica Microsystem S.r.l., Milan, Italy) and Charpy-V impact toughness tests at room temperature and at −22 °C by means of a Charpy Wolpert pendulum (Wolpert Wilson Instruments, Aachen, Germany). In order to confirm the reproducibility of the results, for each heat treatments combination, five micro hardness and three impact toughness measurements were performed. Mechanical properties were linked to the microstructure that was analyzed by a LEICA DMRE optical microscope (Leica Microsystem S.r.l., Milan, Italy) and a Leica Cambridge Stereoscan LEO 440 scanning electron microscope (Leica Microsystem S.r.l., Milan, Italy). For the microstructural analysis, the samples were cut and mounted in phenolic resin, and then polished with standard metallographic preparation technique (grinding with abrasive papers 320, 500, 800, and 1200 grit, followed by polishing with clothes and diamond suspensions 6 μm and 1 μm). After the polishing step, the samples were etched with Nital 5% reagent immerging the sample for 5 s. Both the observations with Optical Microscope (OM) and Scanning Electron Microscope (SEM) were performed on the etched samples. For SEM analysis secondary electron mode was employed. The microstructure of the samples after different tempering treatment was compared with the one of the quenched sample and the one of the initial forging bush after heat treatment. Both micro-hardness and microstructural analysis were performed on Charpy specimens in regions of the samples far from the V notch. In order to investigate the possible presence of retained austenite, X-ray diffraction (XRD) measurements were performed. XRD analysis was carried out using a Siemens D500 diffractometer (Siemens, Munich, Germany) equipped with Cu tube radiation and a graphite-monochromator on the detector side with step size ($2\theta$) of 0.05° and counting time of 5 s/step. The effect of possible texture was eliminated by a rotating fixture on the goniometer. To correlate more deeply the microstructure with mechanical properties, transmission electron microscope analysis was performed as well. The observations were performed with a JEOL 200CX transmission electron microscopy (JEOL Ltd., Tokyo, Japan). Sample preparation for transmission electron microscopy consisted in mechanical thinning to 50 μm, cutting of 3 mm diameter disk and perforation with Twin-Jet polishing technique, using a solution of 95% acetic acid and 5% perchloric acid as electrolyte at 50 V and room temperature.

## 3. Results

### 3.1. Micro-Hardness and Impact Toughness Evaluation

The results of the characterization on the different samples in terms of micro-hardness and impact toughness at room temperature and at −22 °C are summarized in Table 2.

**Table 2.** Micro-Hardness and Impact Toughness of the forged S690 steel after different heat treatments.

| Treatment | Micro-Hardness $HV_{0.2}$ | Impact Toughness Room T (J) | Impact Toughness −22 °C (J) |
|---|---|---|---|
| Water Quenched (after austenitization 890 °C 4 h) | 400 ± 5 | - | - |
| Quenched and Tempered 590 °C 6 h | 276 ± 5 | 220 ± 10 | 48 ± 10 |
| Quenched and Tempered 590 °C 10 h | 275 ± 6 | 237 ± 9 | 60 ± 8 |
| Quenched and Tempered 590 °C 6 h + second tempering 590 °C 4 h | 272 ± 6 | 282 ± 9 | 128 ± 10 |

From the previously reported data, it can be clearly observed that the different tempering treatments do not affect the hardness of the steel. In fact, this property does not vary between the sample quenched and tempered and the samples with the second tempering treatment. Clearly, a reduction from the hardness of the quenched sample was observed for all the tempered samples.

Considering the impact toughness, more significant variations between the various samples can be noticed. In the tests at room temperature, an increase of about 60 J in impact toughness can be observed, passing from the quenched and tempered samples (6 h or 10 h) to the sample with the second tempering treatment. In detail, the sample that showed the higher value of impact toughness is the sample treated with the second tempering at 590 °C for 4 h that exhibits an impact toughness about 60 J higher than the sample quenched and tempered in one step. The difference in the impact toughness between the samples with and without the second tempering treatment is even higher at −22 °C. In this case, the impact toughness raises from 48 J (60 J) of the samples quenched and tempered for 6 h (10 h), to 128 J of the samples which also underwent the double tempering (6 + 4 h) heat treatment sequence.

## 3.2. Microstructural Characterization

In order to correlate the evidences regarding the impact toughness with microstructural changes OM, SEM and TEM observations were performed on all the tempered samples. For comparison the OM microstructures of the initial forging bush after quenching and tempering (Figure 3A) and the test ring samples after quenching (Figure 3B) are also reported.

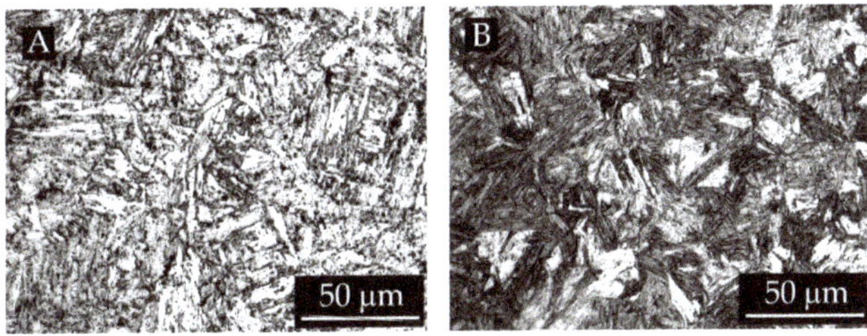

**Figure 3.** OM micrographs of the initial forging bush after austenitization at 890 °C for 4 h, water quenching and tempering at 620 °C for 6 h (**A**) and of test ring samples after austenitization at 890 °C for 4 h and water quenching in laboratory scale furnace (**B**).

From the micrograph reported in Figure 3A the expected microstructure composed by bainite and tempered martensite in the initial forging can be observed. After quenching (Figure 3B), the microstructure is composed by lath martensite and mixed bainite.

After water quenching, the test ring slices underwent different tempering treatments, as previously described, the microstructure of which is shown in Figure 4.

Comparing OM images of Figure 4, it appears that there is no significant microstructural difference among the samples. In fact, microstructure of all samples is substantially composed mainly by bainite with the presence also of low amount of tempered martensite.

In order to investigate in deeper detail possible microstructural differences that can explain the obtained improvement in impact toughness, SEM observations were performed on the three samples, whose results are reported in Figures 5–7.

From the SEM images, the mainly bainitic microstructure of all the samples is confirmed. In all the samples it can be also observed the presence of small carbides (white spots) located mainly along the prior austenitic grain boundaries and in-between bainitic plates. Comparing micrographs of Figures 5–7, no significant differences in the microstructure can be observed even in at higher magnifications.

One of the microstructural changes that can affect impact toughness in this steel, as previously described, is the presence of retained austenite, that is not visible in the OM images. To confirm the absence of retained austenite, XRD analyses were performed and the results can be observed in Figure 8.

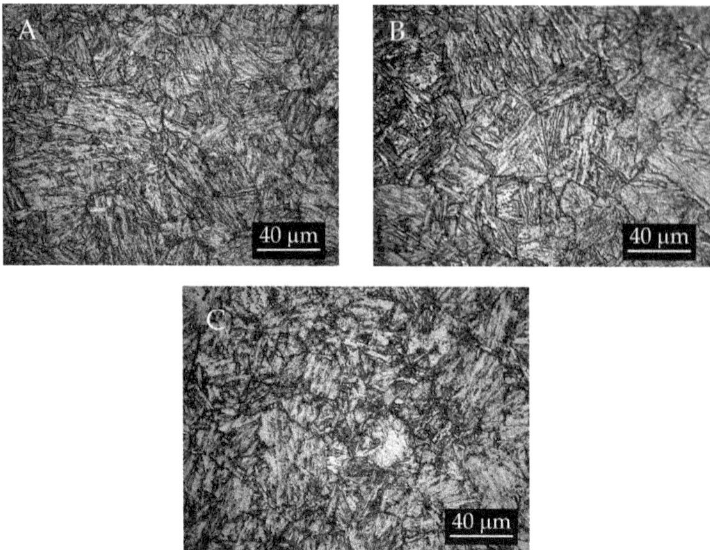

**Figure 4.** OM micrographs of the samples after different heat treatments: quenched and tempered at 590 °C for 6 h (**A**), quenched and tempered at 590°C for 10 h (**B**), quenched and tempered at 590 °C 6 h + 590 °C 4 h (**C**).

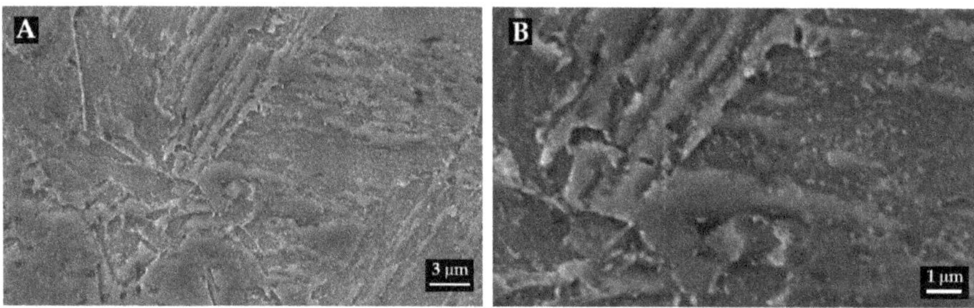

**Figure 5.** SEM micrographs of the sample quenched and tempered at 590 °C for 6 h at low magnification (**A**) and higher magnification (**B**).

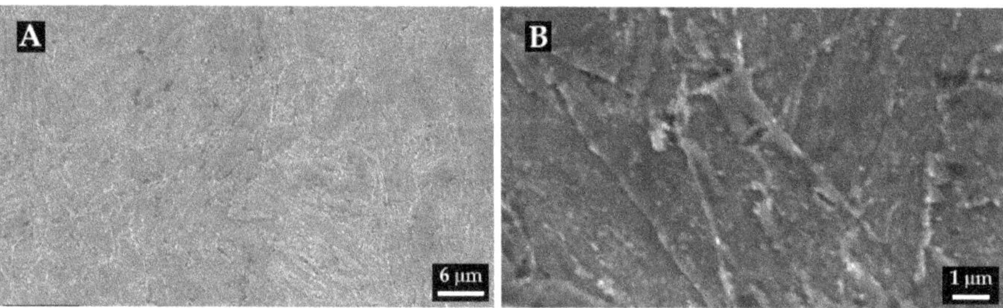

**Figure 6.** SEM micrographs of the sample quenched and tempered at 590 °C for 10 h at low magnification (**A**) and higher magnification (**B**).

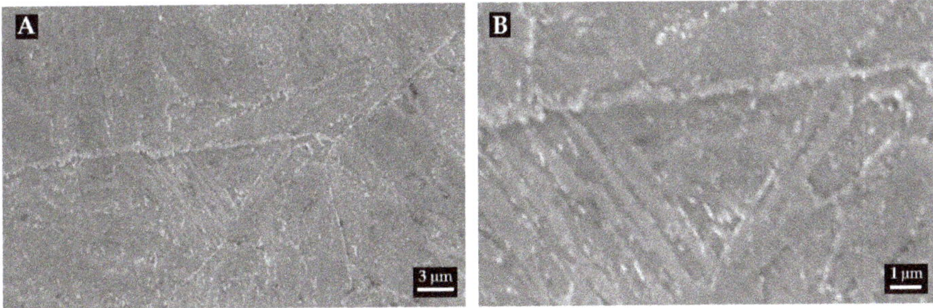

**Figure 7.** SEM micrographs of the sample quenched and tempered at 590 °C for 6 h + 590 °C 4 h at low magnification (**A**) and higher magnification (**B**).

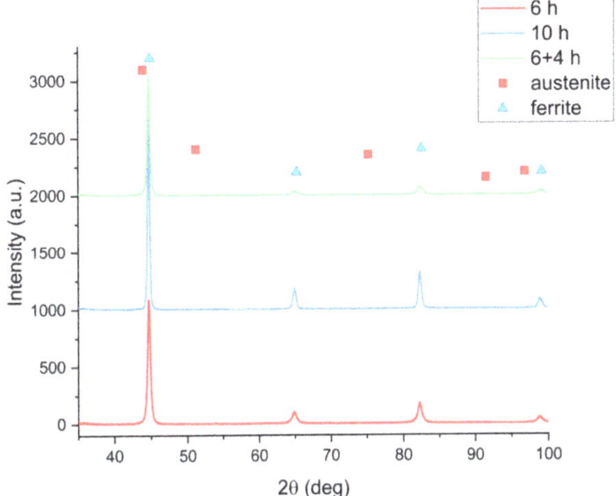

**Figure 8.** XRD analysis of the samples after different heat treatments: quenched and tempered at 590 °C for 6 h (red line); quenched and tempered at 590 °C for 10 h (blue line); quenched and tempered at 590 °C 6 h + 590 °C 4 h (green line).

From XRD analysis, retained austenite was not detected in any of the samples. In fact, only the peaks pertaining to ferrite can be noticed in the XRD patterns.

The shape and distribution of carbides can significantly affect the impact toughness. However, due to the small size of carbides, from SEM observation it was not possible to analyze and compare their shapes. In order to further investigate these microstructural features, TEM analysis on the three samples were performed and the results are reported in Figure 9. The images confirm the mainly bainitic microstructure with the presence of mixed upper and lower bainite, with carbides that can be found both inside and along the bainitic plates in addition to the one's observable at the grain boundaries. Significant variations in the shape of the carbides can be noticed. In detail, in the sample quenched and tempered at 590 °C for 6 h the carbides are lenticular-shaped, as shown in Figure 9A (red circles). Increasing the tempering time to 10 h (Figure 9B) produces an increase in the carbides average dimension, however, keeping an almost lenticular shape. With double tempering 6 h + 4 h (Figure 9C) a more globular shape can be noticed. The evolution in the shape of the carbides (i.e., from lenticular to globular) produces the observed increase in impact toughness. However, the hardness remains constant, despite the increased soaking time or number of steps in the tempering treatment.

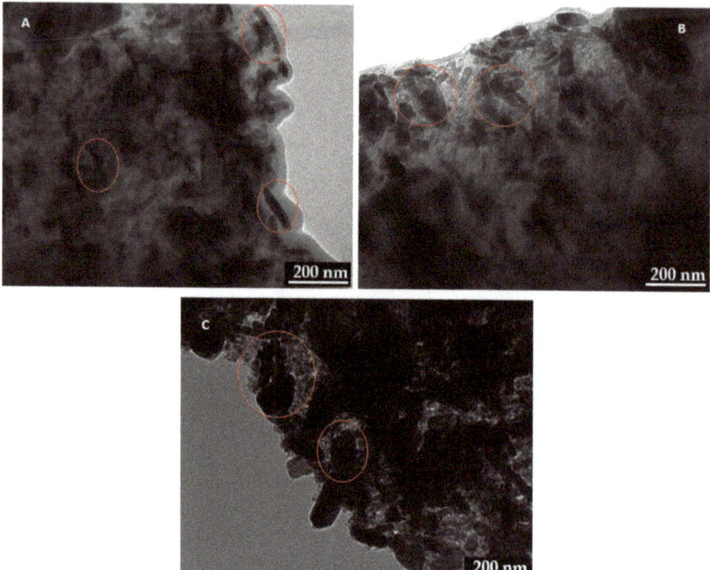

**Figure 9.** TEM micrographs of the samples after different heat treatments: quenched and tempered at 590 °C for 6 h (**A**), quenched and tempered at 590 °C for 10 h (**B**), quenched and tempered at 590 °C 6 h + 590 °C 4 h (**C**).

## 4. Discussion

The results previously reported highlighted that for S690 steel a double tempering of 6 h and 4 h after quenching permits to achieve increased impact toughness if compared to a single tempering of 6 h or 10 h. This behavior is not related to the presence of retained austenite (absent according to XRD analysis evaluation). All the samples present a mainly bainitic microstructure with carbides at prior austenitic grain boundaries and in-between bainitic plates. Also, the presence of tempered martensite was observed. The shape of such carbides is the key of the mechanical behavior of the samples: indeed, passing from single to double tempering it was recorded a modification from lenticular to globular shape. This combination of microstructural features causes the increase in impact toughness. This evidence is in accordance with literature results, obtained on different steels. In detail, Takebayashi et al. [20] found that carbide size and distribution significantly influence impact toughness of martensitic steels.

In this work, the modification in the morphology of carbides is linked with the type of heat treatment: a short soaking time (tempering at 590 °C for 6 h) does not permit the formation of a large number of lenticular carbides. Increasing treatment time (tempering at 590 °C for 10 h) produces an increase in the dimension of carbides but does not substantially modify their shape. Instead, after a double tempering (590 °C 6 h + 4 h) carbides morphology changes, which besides becoming coarser, acquires a spherical shape. However, the hardness of the steel was not affected by such combination of microstructural modifications.

In literature, it is reported that the following carbides, characterized by different thermal stability, can precipitate depending on tempering time and temperature: $M_3C$, $M_2C$, $M_7C_3$, $M_{23}C_6$, and $M_6C$ [11]. Increasing tempering time leads to evolution of the unstable carbides towards the equilibrium ones, whose nature depends on the steel composition. This event is usually accompanied by coarsening and spheroidization of carbides [21].

Carbides sequence precipitation was studied for S690 steel with software simulation in the case of quenching at 890 °C and single tempering at 590 °C, and the results are reported in Figure 10.

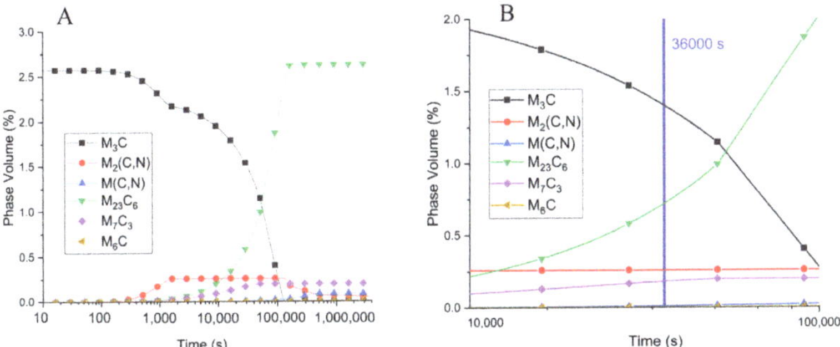

**Figure 10.** Results of the simulation of carbides precipitation during Quenching at 890°C and Temping at 590°C for S690 steel performed with JMatPro® software; overview (**A**) and close up of the region near 10 h treatment time (**B**).

The diagram reported in Figure 10 confirmed, also for S690 steel, the results found in literature: the $M_3C$ (characterized by lenticular morphology) is the first carbide to form and $M_{23}C_6$ (characterized by rod-like morphology) is the equilibrium carbide at this temperature.

In this work, it is likely that double tempering accelerates the transition from metastable to equilibrium carbides, by modifying the mechanism of carbon diffusion as well as diffusion of other elements.

Moreover, the evolution of carbide shape from lenticular to rod-like could explain the observed increase in impact toughness, as also observed by Dudova et al. [22] and Hao et al. [23]. This shape evolution is probably predominant in comparison to coarsening and this explains the almost constant hardness after the different treatments.

## 5. Conclusions

In this work, the effect of different heat treatments on a forged S690 steel was studied. Three different tempering treatments, performed after quenching, were studied: 590 °C for 6 h, 590 °C for 10 h and a double tempering at 590 °C for 6 h + 4 h. The different tempering treatments produce significant variations in impact toughness: the double tempering treatment results in a remarkable improvement in impact toughness of approximately 80 J, in comparison with the sample tempered for 6 h, and of about 60 J in comparison with the single 10 h tempering treatment. None of the samples contain retained austenite, as evidenced by XRD analysis, and are characterized by a mainly bainitic microstructure with the presence of carbides, as observed with OM, SEM, and TEM. The main difference among the various samples is due to the shape of carbides, as evidenced by TEM observation. In detail, with a two-stage tempering treatment, a morphological change was observed from lenticular to globular. These variations can be explained with the modification in the mechanism of transition from metastable to equilibrium carbides during the double tempering. In fact, the formation of the equilibrium carbides induces their spheroidization and can justify the recorded increase in the impact toughness.

**Author Contributions:** Conceptualization, M.T.(Michele Toldo), M.T.( Mario Toniolo), F.S.; methodology, A.Z., K.B.; software, C.G.; validation, M.T.(Michele Toldo), C.G..; formal analysis, D.C.; investigation, L.P., D.C., C.G.; resources, M.T.( Michele Toldo), F.S., M.D.; data curation, L.P., C.G.; writing—original draft preparation, L.P.; writing—review and editing, K.B., A.Z. and C.G.; visualization, C.G.; supervision, K.B., A.Z., M.D.; project administration, M.D.; funding acquisition, M.D. All authors have read and agreed to the published version of the manuscript.

**Funding:** This research received funding from the research contract between FOC Ciscato S.p.a. and the Department of Industrial Engineering, University of Padova.

**Acknowledgments:** The authors want to thank FOC Ciscato S.p.a. for providing the materials and the financial support to the study.

**Conflicts of Interest:** The authors declare no conflict of interest. The funders had no role in the design of the study; in the collection, analyses, or interpretation of data; in the writing of the manuscript, or in the decision to publish the results.

## References

1. Qiang, X.; Bijlaard, F.; Kolstein, H. Dependence of mechanical properties of high strength steel S690 on elevated temperatures. *Constr. Build. Mater.* **2012**, *30*, 73–79. [CrossRef]

2. De Jesus, A.M.P.; Matos, R.; Fontoura, B.F.C.; Rebelo, C.; Simões da Silva, L.; Veljkovic, M. A comparison of the fatigue behavior between S355 and S690 steel grades. *J. Constr. Steel Res.* **2012**, *79*, 140–150. [CrossRef]

3. Di Schino, A.; Di Nunzio, P. Metallurgical aspects related to contact fatigue phenomena in steels for back-up rolls. *Acta Metall. Slovaca* **2017**, *23*, 62. [CrossRef]

4. Di Schino, A.; Di Nunzio, P.E.; Lopez Turconi, G. Microstructure Evolution during Tempering of Martensite in a Medium-C Steel. *Mater. Sci. Forum* **2007**, *558*, 1435–1441. [CrossRef]

5. Di Schino, A. Analysis of heat treatment effect on microstructural features evolution in a micro-alloyed martensitic steel. *Acta Metall. Slovaca* **2016**, *22*, 266. [CrossRef]

6. Gennari, C.; Pezzato, L.; Piva, E.; Gobbo, R.; Calliari, I. Influence of small amount and different morphology of secondary phases on impact toughness of UNS S32205 Duplex Stainless Steel. *Mater. Sci. Eng. A* **2018**, *729*, 149–156. [CrossRef]

7. Ćwiek, J.; Łabanowski, J.; Topolska, S. The effect of long-term service at elevated temperatures on structure and mechanical properties of Cr-Mo-V steel. *Arch. Mater. Sci. Eng.* **2011**, *49*, 33–39.

8. Zielinski, M.A.; Dobrzanski, J.; Golanski, G. Estimation of the residual life of L17HMF cast steel elements after long-term service. *J. Achiev. Mater. Manuf. Eng.* **2009**, *34*, 137–144.

9. Bhadeshia, H.K.D.H. *Bainite in Steels*; Woodhead Pub Ltd.: Cambridge, UK, 2001; Volume 32, ISBN 1861251122.

10. Bhadeshia, H.K.D.H.; Honeycombe, R. *Steels: Microstructure and Properties*; Butterworths-Heinemann (Elsevier): Aalborg, Denmark, 2006; ISBN 9780750680844.

11. Dépinoy, S.; Toffolon-Masclet, C.; Urvoy, S.; Roubaud, J.; Marini, B.; Roch, F.; Kozeschnik, E.; Gourgues-Lorenzon, A.F. Carbide Precipitation in 2.25 Cr-1 Mo Bainitic Steel: Effect of Heating and Isothermal Tempering Conditions. *Metall. Mater. Trans. A Phys. Metall. Mater. Sci.* **2017**. [CrossRef]

12. Krawczyk, J.; Bała, P.; Pacyna, J. The effect of carbide precipitate morphology on fracture toughness in low-tempered steels containing Ni. *J. Microsc.* **2010**. [CrossRef] [PubMed]

13. Horn, R.M.; Ritchie, R.O. Mechanisms of tempered martensite embrittlement in low alloy steels. *Metall. Trans. A* **1978**. [CrossRef]

14. Yen, H.W.; Chiang, M.H.; Lin, Y.C.; Chen, D.; Huang, C.Y.; Lin, H.C. High-temperature tempered martensite embrittlement in quenched-and-tempered offshore steels. *Metals* **2017**, *7*, 253. [CrossRef]

15. Abbasi, E.; Luo, Q.; Owens, D. Microstructural Characteristics and Mechanical Properties of Low-Alloy, Medium-Carbon Steels after Multiple Tempering. *Acta Metall. Sin.* **2019**, *32*, 74–88. [CrossRef]

16. Bakhsheshi-Rad, H.R.; Monshi, A.; Monajatizadeh, H.; Idris, M.H.; Abdul Kadir, M.R.; Jafari, H. Effect of Multi-Step Tempering on Retained Austenite and Mechanical Properties of Low Alloy Steel. *J. Iron Steel Res. Int.* **2011**, *18*, 49–56. [CrossRef]

17. Wang, J.; Kang, Y.; Yu, H.; Ge, W. Effect of Quenching Temperature on Microstructure and Mechanical Properties of Q1030 Steel. *Mater. Sci. Appl.* **2019**, *10*, 665–675. [CrossRef]

18. Dong, D.; Li, H.; Shan, K.; Jia, X.; Li, L. Effects of Different Heat Treatment Process on Mechanical Properties and Microstructure of Q690 Steel Plate. *IOP Conf. Ser. Mater. Sci. Eng.* **2018**, *394*, 022017. [CrossRef]

19. Qin, S.; Song, R.; Xiong, W.; Liu, Z.; Wang, Z.; Guo, K. Microstructure evolution and mechanical properties of grade E690 offshore platform steel. In *HSLA Steels 2015, Microalloying 2015 & Offshore Engineering Steels 2015*; John Wiley & Sons, Inc.: Hoboken, NJ, USA, 2015; Volume 2, pp. 1117–1123, ISBN 9781119223399.

20. Takebayashi, S.; Ushioda, K.; Yoshinaga, N.; Ogata, S. Effect of carbide size distribution on the impact toughness of tempered martensitic steels with two different prior austenite grain sizes evaluated by instrumented charpy test. *Mater. Trans.* **2013**. [CrossRef]

21. Li, Z.; Jia, P.; Liu, Y.; Qi, H. Carbide Precipitation, Dissolution, and Coarsening in G18CrMo2–6 Steel. *Metals* **2019**, *9*, 916. [CrossRef]

22. Dudova, N.; Mishnev, R.; Kaibyshev, R. Effect of tempering on microstructure and mechanical properties of boron containing 10%Cr steel. *ISIJ Int.* **2011**. [CrossRef]

23. Hao, X.; Gao, M.; Zhang, L.; Zhao, X.; Liu, K. Microstructure of annealed 12Cr13 stainless steel and its effect on the impact toughness. *Jinshu Xuebao* **2011**. [CrossRef]

*Article*

# Mechanical Properties and Microstructure Characterization of AISI "D2" and "O1" Cold Work Tool Steels

**Mohammed Algarni**

Mechanical Engineering Department, Faculty of Engineering, King Abdulaziz University, P.O. box 344, Rabigh 21911, Saudi Arabia; malgarni1@kau.edu.sa

Received: 25 September 2019; Accepted: 28 October 2019; Published: 30 October 2019

**Abstract:** This research analyzes the mechanical properties and fracture behavior of two cold work tool steels: AISI "D2" and "O1". Tool steels are an economical and efficient solution for manufacturers due to their superior mechanical properties. Demand for tool steels is increasing yearly due to the growth in transportation production around the world. Nevertheless, AISI "D2" and "O1" (locally made) tool steels behave differently due to the varying content of their alloying elements. There is also a lack of information regarding their mechanical properties and behavior. Therefore, this study aimed to investigate the plasticity and ductile fracture behavior of "D2" and "O1" via several experimental tests. The tool steels' behavior under monotonic quasi-static tensile and compression tests was analyzed. The results of the experimental work showed different plasticity behavior and ductile fracture among the two tool steels. Before fracture, clear necking appeared on "O1" tool steel, whereas no signs of necking occurred on "D2" tool steel. In addition, the fracture surface of "O1" tool steel showed cup–cone fracture mode, and "D2" tool steel showed a flat surface fracture mode. The dimple-like structures in scanning electron microscope (SEM) images revealed that both tool steels had a ductile fracture mode.

**Keywords:** fracture; cold work tool steel; AISI D2; AISI O1; high carbon steel

## 1. Introduction

In the machining and forming industry, tool steels were invented to increase manufacturing economic efficiency due to their enhanced mechanical properties, such as high strength, wear resistance, hardness, and toughness. Metal machining and forming are essential for metal part production in many industries. The automotive industry, for example, has experienced an increase in car production, which has led to an increase in the demand for tool steels. The same increase in demand has also been experienced in other industries, such as aerospace, transport, and precision industries. The vast increase in production has resulted in substantial growth in the metal forming industry, at a rate of 3% to the year 2019 [1]. Tool steels are categorized into six classes: cold work, hot work, shock resisting, mold, high speed, and special-purpose tool steels [2]. The most important class of all is cold work tool steels. This research investigated two types of cold work tool steel that have a high content of carbon: "D2" and "O1". These metals are used for many types of cutting and punching tools and dies and many other applications. They have high hardness, high wear resistance, and are inexpensive [3–5]. Previous studies have tested the two metals to determine their wearing properties and resistance, with "O1" tool steel being found to have excellent machinability, whereas "D2" had better wearing resistance [6–8].

The oil-hardening "O1" cold work steel (UNS# T31501) is a low-cost metal that has high hardness and wear resistance due to the high carbon content along with moderate levels of different elements, such as chromium (Cr) and silicon (Si). The high content of Si alloy element increases machinability and die life. In addition, the existence of tungsten (W) alloy element attains high abrasion resistance and highly sharp cutting edges. Thus, this tool steel is used in surface finishing tools and woodworking knives [3,9–11]. The high-carbon, high-chromium AISI "D2" cold work steel (UNS# T30402) is particularly poor in terms of machinability and toughness. In the fabrication process, "D2" is highly resistant to softening and wearing, with minimal microstructure distortion and high resistance to cracking during metal formation and fabrication [12]. Therefore, in long-duration fabrication processes, "D2" is preferable for manufacturers. In addition, "D2" tool steel is heavily used in piercing punches and dies, forging operations, and trimming tools due to its high wear resistance [9,13–15]. Moreover, it is generally known that "D2" tool steel is difficult to weld (nonweldable), and it is particularly hard to attain a high-quality welded joint by conventional welding methods due to its high carbon content and significant amount of carbides. A recent study [16] proposed a novel thixowelding technology for joining "D2" steels with different joining temperatures, holding time, and postweld heat treatment to investigate the joints' mechanical and microstructural properties. The results demonstrated a significant improvement in its tensile strength for heat- vs. non-heat-treated joints. Another study [17] investigated the effect of post-tempering cryogenic treatment on the mechanical properties of "D2" tool steel. The results showed an improvement in fracture toughness, reduction in residual stresses, and no change in hardness and modulus values.

In the present study, the characteristics of mechanical properties are reported. Tensile strength, compression strength, hardness, elongation at fracture, and reduction area at fracture in addition to the plasticity and ductile fracture behavior of two tool steel metals—AISI "D2" and "O1"—under monotonic loading conditions were investigated. Furthermore, fracture surfaces, dimensional stability, and microstructure features were studied. Optical measurements and optical microscopic investigations were also conducted.

## 2. Experimental Procedure

### 2.1. Sample Preparation

Two steel blocks of AISI "D2" and "O1" were purchased from an ASTM-certified local steel shop. The specimens were cut from the two steel blocks and were subjected to hardening and tempering heat treatment process. The tempering temperatures for "D2" and "O1" were 200 and 250 °C, respectively. The fabrication and machining of the specimens were done in two different shapes: (a) smooth round bar (Figure 1) and (b) cylindrical shape (Figure 2). The smooth round bar was designed for tensile tests, whereas the cylindrical specimen was designed for compression tests. The shape and dimensions of both specimens' geometry are shown in mm in Figures 1 and 2. Note that the smooth round bar and cylinder specimens of similar metals were machined from one steel block to ensure similarity in properties. The number of repetitions of each test was two; hence, a total of two tensile tests and two compression tests for each metal were carried out. The chemical compositions of AISI "D2" and "O1" are shown in Table 1. In addition, the critical and austenization temperatures are listed in Table 2.

**Table 1.** Chemical composition of "D2" and "O1" tool steels in mass percent (%).

| Metal Type | C | Mn | Si | Cr | Mo | W | V | Fe |
|---|---|---|---|---|---|---|---|---|
| "D2" | 1.52 | 0.34 | 0.31 | 12.05 | 0.76 | - | 0.92 | Balance. |
| "O1" | 0.94 | 1.2 | 0.32 | 0.52 | - | 0.53 | 0.19 | Balance. |

**Table 2.** Critical and austenization temperatures of "D2" and "O1" tool steels.

| Metal Type | Ac$_1$ | Ac$_3$ | Ar$_1$ | Ar$_3$ | Austenization Temperature |
| --- | --- | --- | --- | --- | --- |
| "D2" | 788 °C | 845 °C | 769 °C | 744 °C | 1010–1024 °C |
| "O1" | 732 °C | 760 °C | 703 °C | 671 °C | 802–816 °C |

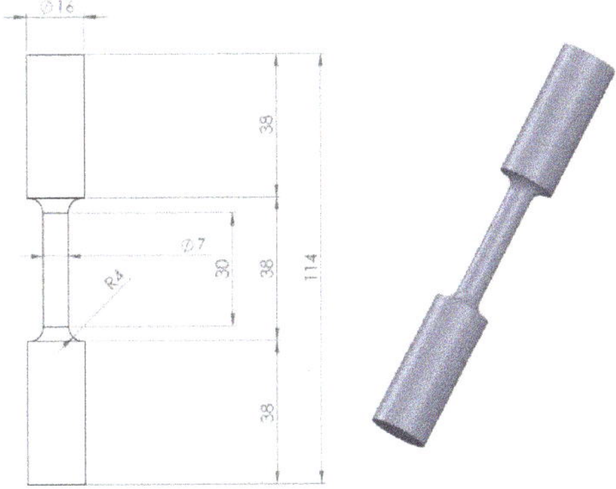

**Figure 1.** Tensile test specimen shape and geometry in mm.

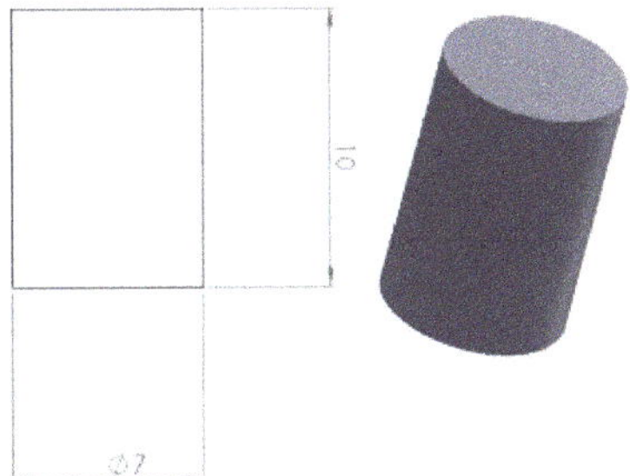

**Figure 2.** Compression test specimen shape and geometry in mm.

## 2.2. Experiments

The load frame used was a servohydrolic testing machine manufactured by MTS systems corporations®in Eden Prairie, MN, USA with a 100 kN loading cell of tension and compression force limit. The tests were conducted at room temperature with a strain rate of 0.005 mm/s. The strain reading was captured and recorded using an optical measurement system termed digital imaging correlation (DIC) [18]. The DIC type was VIC-2D version 5 software made by Correlated Solutions Inc®in Irmo, SC, USA. DIC requires specific preparation (painting the steel specimen) prior to testing in order to provide sufficient contrast for the camera. The specimens were sprayed in white and spackled in black to create a fine contrast for the DIC to capture the strain.

## 3. Results and Discussion

### 3.1. Tensile and Compression Tests

The engineering stress–strain flow performance (total elongation and tensile strength) of "D2" and "O1" subjected to tensile tests at room temperature are shown in Figure 3. The yield strength of "D2" and "O1" and other basic mechanical properties are listed in Table 3. The modulus of toughness (tensile toughness), fracture strength, fracture length, fracture strain, and gauge length are all shown in

Table 4. The differences in content of the alloying elements in "D2" and "O1" tool steels changed the behavior of the stress–strain flow. For example, the higher ductility and toughness of "D2" over "O1" tool steels were due to the high content of molybdenum (Mo), vanadium (V), and Cr. On the other hand, the higher yield tensile strength and ultimate tensile strength (UTS) of "O1" compared to "D2" tool steels were due to the increased content of tungsten and manganese (Mn). The "D2" steel behavior under monotonic loading showed particularly high hardening and substantially low softening due to the high content of Mo and Cr. The range of hardness for "O1" and "D2" steels was 56–58 and 60–62 HRC, respectively. Note that all data reported are the mean value of many testing points for each specimen.

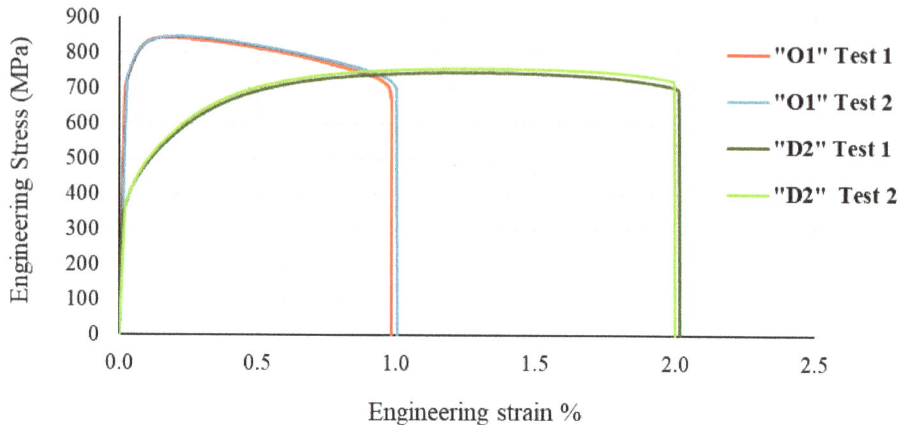


**Figure 3.** Tensile engineering stress–strain curves to fracture for "D2" and "O1" tool steels.

**Table 3.** Basic mechanical properties of "D2" and "O1" tool steels.

| Specimen | Modulus of Elasticity | 0.2% Offset Yield Strength | Yield Strength | UTS |
|---|---|---|---|---|
| AISI "D2" | 203 GPa | 411 MPa | 350 MPa | 758 MPa |
| AISI "O1" | 211 GPa | 829 MPa | 758 MPa | 846 MPa |

**Table 4.** Experimental tensile tests data summary of "D2" and "O1" tool steels.

| Specimen | Modulus of Toughness | Fracture Strength | Displacement at Fracture | Gauge Length | Fracture Strain | Area Reduction |
|---|---|---|---|---|---|---|
| AISI "D2" | 81 MPa | 723 MPa | 0.61 mm | 30 mm | 1.97% | 1.3% |
| AISI "O1" | 68 MPa | 703 MPa | 0.35 mm | 30 mm | 1.09% | 19.7% |

In contrast, the "O1" steel behavior under monotonic loading showed a highly narrow strain range during hardening (before UTS) and a higher range of strain in softening (beyond UTS). The high amount of metal softening during the tensile strength was seen during the experimental test in the form of necking before fracture. The compression stress–strain flow is shown in Figure 4. The cylinder specimens were compressed to approximately −90 kN (load cell maximum capacity is ±100 kN) at a strain rate of 0.005 mm/s while the strain flow was captured. The modulus of elasticity and compression yield of "O1" steel was higher than "D2" steel. The compression plasticity flow, shown in Figure 4, increased as the stresses increased due to the geometry change in the cylinder specimen.

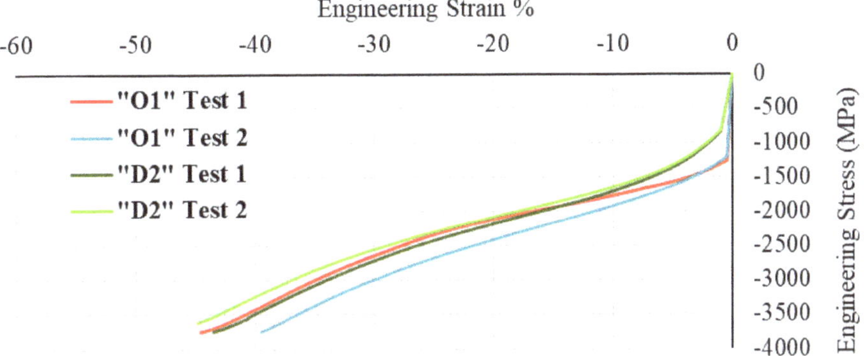

**Figure 4.** Compression engineering stress–strain curves for "D2" and "O1" tool steels.

Another observation is related to the necking behavior of both tool steels. The significant necking before fracture (Figure 5) of "O1" steel was represented in the form of softening beyond the UTS. The calculated area reduction at fracture was almost 20%. However, "D2" tool steel showed almost no necking prior to fracture (Figure 6).

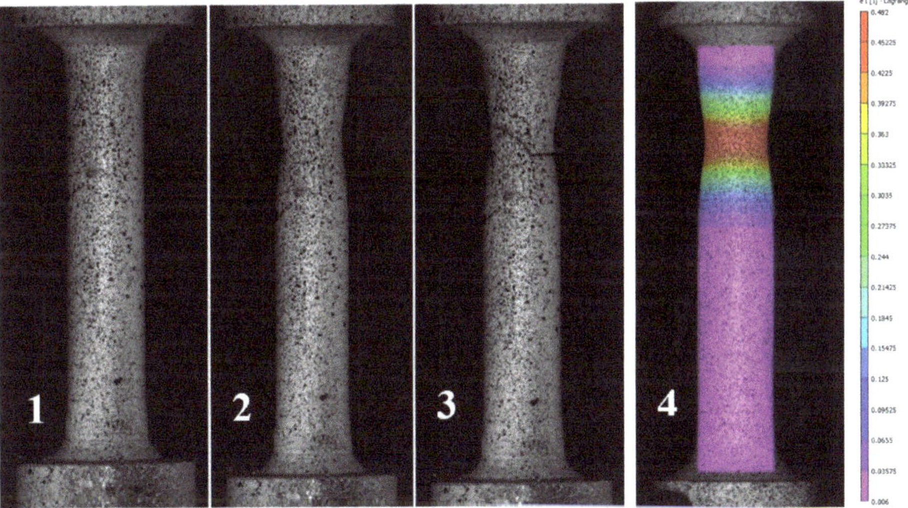

**Figure 5.** "O1" specimen under tension test. (**1**) Specimen before testing, (**2**) necking prior to fracture, (**3**) specimen post fracture, and (**4**) contour plot showing the Lagrange strain localization before fracture in red (maximum stain) and purple (minimum strain).

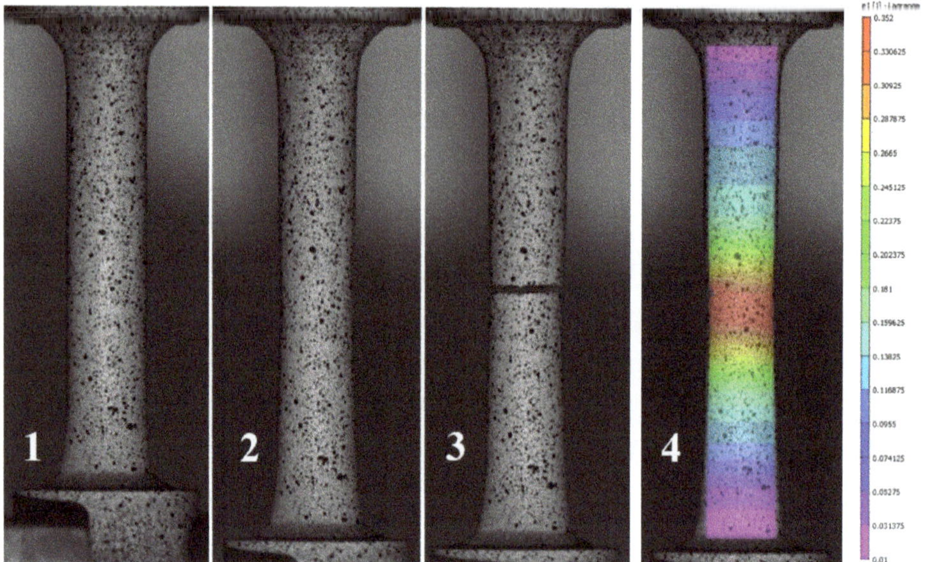

**Figure 6.** "D2" specimen under tension test. (**1**) Specimen before testing, (**2**) no necking prior to fracture, (**3**) specimen post fracture, and (**4**) the contour plot showing the Lagrange strain localization before fracture in red (maximum stain) and purple (minimum strain).

### 3.2. Fracture of Specimens

The "D2" tool steel specimen subjected to tensile loadings is shown in Figure 6. The images depict the sequential deformation process as a result of the tensile loading process. The testing specimen setting was set initially as in Figure 6(1). This image can be used as a reference for comparison. The maximum elongation is shown in Figure 6(2), where necking prior to fracture can hardly be seen. Based on the stress–strain curve of "O1" tool steel, shown in Figure 3, this type showed almost no softening behavior post UTS point, which explains the negligible necking behavior during the experiment. The crack initiated and instantly propagated toward the outer radius, similar to the "O1" tool steel specimen (Figure 6(3)). The strain measurements an instant before fracture are shown in Figure 6(4) with the use of DIC. DIC can also predict the crack initiation location. The crack initiation location and propagation path prediction by DIC have previously been investigated and proven in many studies. [19–22]. In the case of "O1" tool steel, Figure 6(4) shows high strain concentrations (in red) that resulted in metal cracks and fracture (Figure 6(3)). The fracture location prediction agreed with the experimental results. Finally, the failure mode showed a flat fracture surface (Figure 7). It is recommended that the reader refer to [23,24] in order to understand why the fracture surface shape differs from one metal to another.

Similarly, the "O1" tool steel specimen subjected to tensile tests is shown in Figure 5. The collection of images in Figure 5 shows the deformation sequence during the loading process. Figure 5(1) shows the specimen prepared for testing before any loading was applied. This figure can be used for comparison and reference reasons. Figure 5(2) shows the specimen at its highest elongation capacity without fracturing. This moment records the maximum necking (localized area reduction) of the specimen. The necking occurred just before the crack initiated and instantly propagated toward the outer radius, causing full specimen fracture (Figure 5(3)). The color contour plot in Figure 5(4) shows the highest accumulation strain by the DIC just before the fracture occurred. The location of the highest accumulation strain is at the center of the necking area, colored in red. The crack initiation and propagation that appear in Figure 5(3) coincide with the high strain's measurement location developed due to tensile loading in Figure 5(4). In both metals, the crack location prediction by DIC was in good agreement with the experiment results. Finally, the failure mode of the fracture surface on "O1" tool steel under tensile loading showed a cup–cone-like fracture pattern (Figure 8).

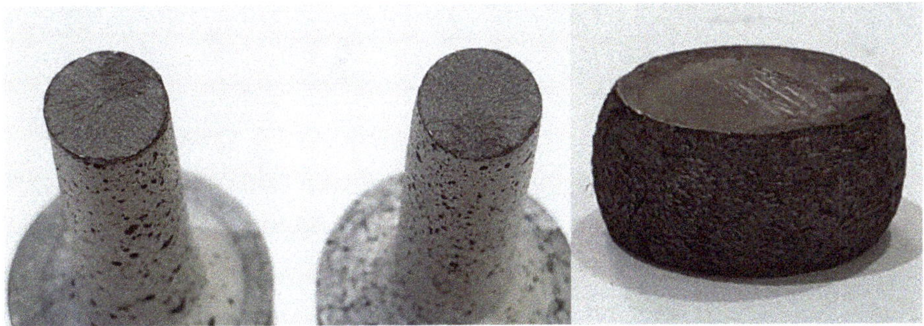

**Figure 7.** A flat fracture mode surface for "D2" tool steel under tension (**left**) and flattened shape of the cylindrical specimen (**right**) under compression.

**Figure 8.** A cup–cone mode fracture surface and clear necking for "O1" tool steel under tension (**left**) and flattened shape of the cylindrical specimen (**right**) under compression.

### 3.3. Microstructure

The microstructure on the fractured surfaces of the "O1" specimens with a cup–cone shape and the "D2" specimens with a flat shape were analyzed with different magnifications using a scanning electron microscope (SEM) type JSM-7610FPlus Schottky Field Emission made by JOEL ltd. (Tokyo, Japan). SEM analysis assists in determining the fracture mode for both tool steels [25]. The existence of the parabolic dimple-like structures in the SEM images revealed that both tool steels had a ductile fracture mode (Figure 9). However, for better analysis to assess the fracture mode and failure mechanism, in-situ X-ray tomography can be performed [26–28].

A careful inspection of both tool steels showed some small differences. The surface fracture morphology was rougher on the "D2" fractured surface compared to the "O1" fractured surface. However, the average size of microvoids in the "O1" specimens was smaller compared to the features in the "D2" specimens (Figure 10). Note that the average microvoid size increased as the hardness decreased for both steels, as can be seen on the "O1" and "D2" fractured surfaces in Figure 11. The SEM micrographs showed different microstructures when the two steel metals were compared. The microvoids were deeper on "O1" and sharper on "D2". The fracture surfaces of "O1" had smaller dimples with fewer cleavage planes compared to "D2" (Figure 11). This observation reasonably explains the higher elongation in "D2" tool steel specimens.

**Figure 9.** SEM morphologies of fracture surfaces of "D2" (**left**) and "O1" (**right**) with ×1500 magnification.

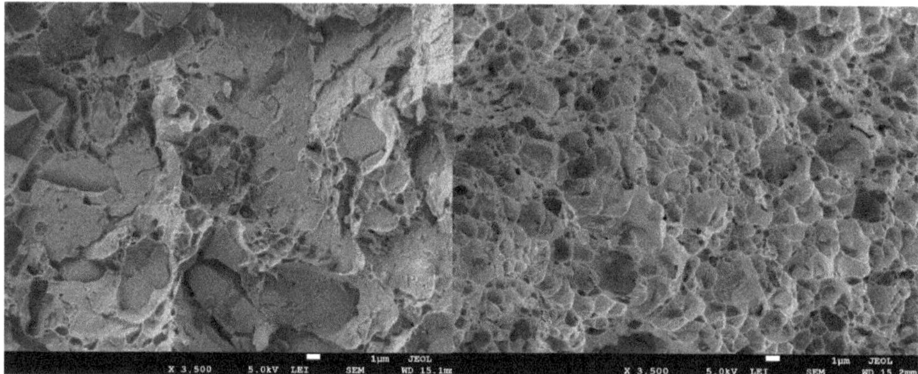

**Figure 10.** SEM morphologies of fracture surfaces of "D2" (**left**) and "O1" (**right**) with ×3500 magnification.

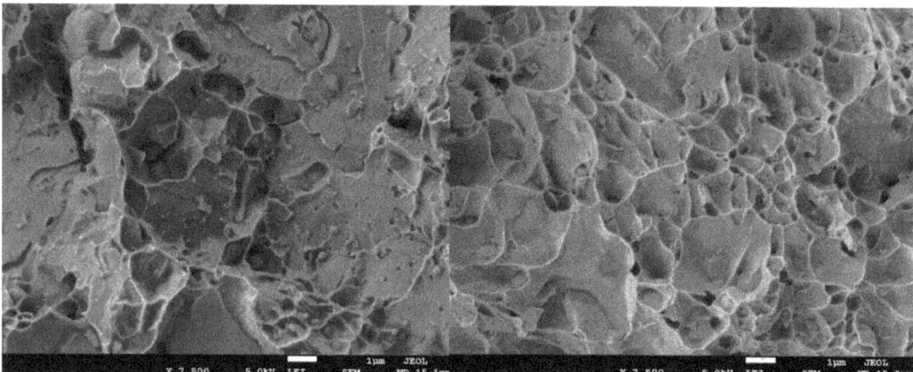

**Figure 11.** SEM morphologies of fracture surfaces of "D2" (**left**) and "O1" (**right**) with ×7500 magnification.

## 4. Conclusions

In this research study, the plasticity and fracture behavior of two tool steels—AISI "D2" and "O1"—were investigated. The tool steels were purchased and fabricated locally, and the specimens were designed for two mechanical tests: tensile and compression. The results demonstrate that AISI "D2" is recommended for applications that require moderate toughness and dimensional stability. In contrast, AISI "O1" is more suitable for applications that require better machinability performance and an excellent combination of high hardness and toughness. The following points conclude the results of the research:

1. The tensile yield strength of "O1" tool steel was higher than "D2" tool steel.
2. The specimens of "O1" tool steel showed vivid necking prior to fracture with 19.7% area reduction, whereas the specimens of "D2" tool steel demonstrated no necking throughout the loading process (1.3% area reduction).
3. The compression yield strength was higher for "O1" than for "D2" tool steel.
4. The surface fracture for "O1" was cup–cone, whereas it was flat for "D2" tool steel.
5. DIC was used to measure surface strains and predict cracks initiation location. The high localized strains identified in the DIC images pointed out where the cracks initiated. The crack initiation prediction was in good agreement with the results of the experiments for both tool steel types.
6. The parabolic dimple-like structures in the SEM images revealed that both tool steels had a ductile fracture mode.
7. The SEM images showed deeper microvoids on "O1" and sharper ones on "D2". The fracture surfaces of "O1" had smaller dimples with less cleavage planes compared to "D2".

**Funding:** This project was funded by the Deanship of Scientific Research (DSR), King Abdulaziz University, Jeddah, under grant No. (DF-309-829-1441). The author, therefore, gratefully acknowledges the DSR technical and financial support.

**Conflicts of Interest:** The authors declare no conflict of interest.

## References

1. Toboła, D.; Brostow, W.; Czechowski, K.; Rusek, P. Improvement of wear resistance of some cold working tool steels. *Wear* **2017**, *382*, 29–39. [CrossRef]
2. Budinski, K.G.; Budinski, M.K. *Engineering Materials: Properties and Selection*; Prentice Hall: Upper Saddle River, NJ, USA, 2010; p. 756.
3. Bourithis, L.; Papadimitriou, G.D.; Sideris, J. Comparison of wear properties of tool steels AISI D2 and O1 with the same hardness. *Tribol. Inter.* **2006**, *39*, 479–489. [CrossRef]
4. Glaeser, W.A. *Characterization of Tribological Materials*; Momentum Press: New York, NY, USA, 2012.
5. Kheirandish, S.; Saghafian, H.; Hedjazi, J.; Momeni, M. Effect of heat treatment on microstructure of modified cast AISI D3 cold work tool steel. *J. Iron. Steel Res. Int.* **2010**, *17*, 40–45. [CrossRef]
6. Hutchings, I.; Shipway, P. *Tribology: Friction and Wear of Engineering Materials*; Butterworth-Heinemann: Oxford, UK, 2017.
7. Straffelini, G. Materials for Tribology. In *Friction and Wear*; Springer: Berlin, Germany, 2015; pp. 159–199.
8. Lansdown, A.R.; Price, A.L. *Materials to Resist Wear*; Pergamon Press: Elmsford, NY, USA, 1986.
9. Roberts, G.A.; Kennedy, R.; Krauss, G. *Tool Steels*; ASM international: Cleveland, OH, USA, 1998.
10. Kataria, R.; Kumar, J. A comparison of the different multiple response optimization techniques for turning operation of AISI O1 tool steel. *J. Eng. Res.* **2014**, *2*, 1–24. [CrossRef]
11. Camacho, L.D.A.; Miranda, S.G.; Moreno, K.J. Tribological performance of uncoated and TiCN-coated D2, M2 and M4 steels under lubricated condition. *J. Iron. Steel Res. Int.* **2017**, *24*, 823–829. [CrossRef]
12. Wei, M.-x.; Wang, S.-q.; Lan, W.; Cui, X.-h.; Chen, K.-m. Selection of heat treatment process and wear mechanism of high wear resistant cast hot-forging die steel. *J. Iron. Steel Res. Int.* **2012**, *19*, 50–57. [CrossRef]
13. Koshy, P.; Dewes, R.C.; Aspinwall, D.K. High speed end milling of hardened AISI D2 tool steel (~58 HRC). *J. Mater. Process. Technol.* **2002**, *127*, 266–273. [CrossRef]

14. Lima, J.G.; Avila, R.F.; Abrao, A.M.; Faustino, M.; Davim, J.P. Hard turning: AISI 4340 high strength low alloy steel and AISI D2 cold work tool steel. *J. Mater. Process. Technol.* **2005**, *169*, 388–395. [CrossRef]
15. Ma, X.; Liu, R.; Li, D.Y. Abrasive wear behavior of D2 tool steel with respect to load and sliding speed under dry sand/rubber wheel abrasion condition. *Wear* **2000**, *241*, 79–85. [CrossRef]
16. Mohammed, M.N.; Omar, M.Z.; Al-Zubaidi, S.; Alhawari, K.S.; Abdelgnei, M.A. Microstructure and mechanical properties of thixowelded AISI D2 tool steel. *Metals* **2018**, *8*, 316. [CrossRef]
17. Sola, R.; Giovanardi, R.; Parigi, G.; Veronesi, P. A novel method for fracture toughness evaluation of tool steels with post-tempering cryogenic treatment. *Metals* **2017**, *7*, 75. [CrossRef]
18. He, W.; Hayatdavoudi, A. A comprehensive analysis of fracture initiation and propagation in sandstones based on micro-level observation and digital imaging correlation. *J. Pet. Sci. Eng.* **2018**, *164*, 75–86. [CrossRef]
19. Caminero, M.A.; Lopez-Pedrosa, M.; Pinna, C.; Soutis, C. Damage monitoring and analysis of composite laminates with an open hole and adhesively bonded repairs using digital image correlation. *Composites Part B* **2013**, *53*, 76–91. [CrossRef]
20. Caminero, M.A.; Pavlopoulou, S.; Lopez-Pedrosa, M.; Nicolaisson, B.G.; Pinna, C.; Soutis, C. Analysis of adhesively bonded repairs in composites: Damage detection and prognosis. *Compos. Struct.* **2013**, *95*, 500–517. [CrossRef]
21. Caminero, M.A.; Lopez-Pedrosa, M.; Pinna, C.; Soutis, C. Damage assessment of composite structures using digital image correlation. *Appl. Compos. Mater.* **2014**, *21*, 91–106. [CrossRef]
22. Romanowicz, P.J. Experimental and numerical estimation of the damage level in multilayered composite plates. *Materialwiss. Werkstofftech.* **2018**, *49*, 591–605. [CrossRef]
23. Algarni, M.; Bai, Y.; Choi, Y. A study of Inconel 718 dependency on stress triaxiality and Lode angle in plastic deformation and ductile fracture. *Eng. Fract. Mech.* **2015**, *147*, 140–157. [CrossRef]
24. Tvergaard, V.; Needleman, A. Analysis of the cup–cone fracture in a round tensile bar. *Acta Metall.* **1984**, *32*, 157–169. [CrossRef]
25. Lee, W.-S.; Lin, C.-F.; Liu, T.-J. Strain rate dependence of impact properties of sintered 316L stainless steel. *J. Nucl. Mater.* **2006**, *359*, 247–257. [CrossRef]
26. Landron, C. Ductile Damage Characterization in Dual-Phase steels Using X-Ray Tomography. PhD Thesis, The National Institute of Applied Sciences of Lyon, Lyon, France, 21 December 2011.
27. Weck, A.; Wilkinson, D.S.; Maire, E.; Toda, H. Visualization by X-ray tomography of void growth and coalescence leading to fracture in model materials. *Acta Mater.* **2008**, *56*, 2919–2928. [CrossRef]
28. Toda, H.; Maire, E.; Yamauchi, S.; Tsuruta, H.; Hiramatsu, T.; Kobayashi, M. In situ observation of ductile fracture using X-ray tomography technique. *Acta Mater.* **2011**, *59*, 1995–2008. [CrossRef]

MDPI
St. Alban-Anlage 66
4052 Basel
Switzerland
Tel. +41 61 683 77 34
Fax +41 61 302 89 18
www.mdpi.com

*Metals* Editorial Office
E-mail: metals@mdpi.com
www.mdpi.com/journal/metals